FIRST WOMAN

FIRST WOMAN

Joanne Simpson and the Tropical Atmosphere

James Rodger Fleming

OXFORD
UNIVERSITY PRESS

OXFORD

UNIVERSITY PRESS

Great Clarendon Street, Oxford, OX2 6DP,
United Kingdom

Oxford University Press is a department of the University of Oxford.
It furthers the University's objective of excellence in research, scholarship,
and education by publishing worldwide. Oxford is a registered trade mark of
Oxford University Press in the UK and in certain other countries

Published in the United States of America by Oxford University Press
198 Madison Avenue, New York, NY 10016, United States of America

British Library Cataloguing in Publication Data

Data available

Library of Congress Control Number: 2020940485

ISBN 978–0–19–886273–4

Printed and bound by
CPI Group (UK) Ltd, Croydon, CR0 4YY

Acknowledgments

This book could not have been written without the aid of archivists, librarians, and staff at the Arthur and Elizabeth Schlesinger Library at Harvard University, the University of Chicago Library, Marine Biological Laboratory/Woods Hole Oceanographic Institute Library and Archives, and Colby College Library. Special thanks to Marilyn Dunn, The Lia Gelin Poorvu Executive Director of the Arthur and Elizabeth Schlesinger Library on the History of Women in America and Librarian of the Radcliffe Institute; Catherine Uecker, Head of Research and Instruction, Special Collections Research Center, University of Chicago Library; Jennifer Walton, Director of Library Services at the Marine Biological Laboratory Library and Co-Director of the MBLWHOI Library, and Kara Kugelmeyer, Head of Research and Instruction at Colby College.

Student research assistants, Abe Krieger, Sophie Swetz, Ava Colarusso, Maya Meltsner, Carol Lipshultz, Talia Gebhard, Anna Chen, and Heather Jahrling worked diligently through several drafts, responded to research questions, gave several very public presentations, and read the manuscript with me many times.

I thank my faculty colleagues for their valuable assistance in preparing this book. Dan Cohen, Sarah Duff, Mary Ellis Gibson, Charles Orzech, and Christopher Walker read an early draft and provided valuable feedback over wine, cheese, and typos. Lijing Jiang and Elizabeth McGrath asked probing questions that helped sharpen the narrative. Colleagues working in Europe: Katherine Anderson, Michael Börngen, Cornelia Lüdecke, Hans Volkert, and Magnus Volsett helped immensely to trace early women with advanced training in meteorology. Claire Parkinson at NASA Goddard served as the host for a first round of informal interviews with Patricia Aldridge, Scott Braun, Bob Cahalan, Franco Einaudi, Ida Hakkarinen, Gerry Heymsfield, Holly McIntyre-Dewitt, William Lau, Andrew Negri, Kelly Pecnick, Wei-Kuo Tao, Anne Thompson, and Dorothy Zukor. Thanks to Claire for her hospitality, and thanks to all for their input and insights. The NASA History Office: Bill Berry, Colin Fries, Steve Garber, and Elizabeth Suckow, and the NOAA Central Library, especially Skip Theberge, provided key documents. Colleagues Anjuli Bamzai, Kerry Emmanuel, Roger Launius, Roger Pielke, Sr., and Margaret Weitecamp also added important and varied perspectives.

Video and audio interviews with Rick Anthes, Bill Cotton, Jeff Halverson, Kristina Katsaros, Chris Kummerow, Peggy LeMone, Steven Rutledge, Louis Uccellini, and Warren Washington appear in short video footnotes that are posted on a YouTube playlist (Joanne Simpson Vignettes). These video clips constitute important enhancements to the text as well as a new style of conducting oral interviews. A special thanks to Joanne's children, Steven Malkus and Karen E. Malkus-Benjamin, for sharing their memories

Talks presented at Colorado State University, the History of Science Society, the Society for the History of Technology, the University of Maryland Baltimore County, NASA Goddard, the American Meteorological Society, and Colby College helped establish the narrative and provided opportunities for valuable feedback. Several anonymous reviews from the press clarified important issues.

Finally, a special acknowledgment to Francesca McMahon and Sinduja Abirami, whose efficiency, thoroughness, and gentle editorial touch dazzled me.

Contents

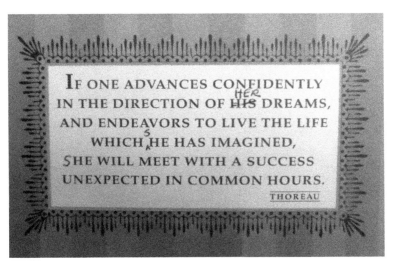

IF ONE ADVANCES CONFIDENTLY
IN THE DIRECTION OF ~~HIS~~ HER DREAMS,
AND ENDEAVORS TO LIVE THE LIFE
WHICH ꜱHE HAS IMAGINED,
ꜱHE WILL MEET WITH A SUCCESS
UNEXPECTED IN COMMON HOURS.

THOREAU

Sɪᴍᴘsᴏɴ Pᴀᴘᴇʀs, 2.7. Original in color.

Introduction

Joanne Simpson transformed the science of the tropical atmosphere and set a course in science for professional women to follow. She had a lifelong passion for clouds and severe storms, flying into and above them, measuring and modeling them, theorizing about the role of tropical clouds in the planetary circulation, and mentoring a generation of tropical meteorologists. In 1993, just shy of her seventieth birthday, Joanne commandeered a fully equipped NASA-owned DC8 research airplane during a field project to study El Niño, and flew several flights directly into tropical cyclone Oliver in the Coral Sea, some 500 km off the coast of Townsville, Australia. She and the crew did this on several consecutive days. The aircraft was equipped with radar being tested for use on a new satellite to measure tropical rainfall, and they wanted to use it to collect the best possible data on storm structure and dynamics. The third flight, directly into the storm, pushed the plane to its limits. The excessive humidity and turbulent shaking shorted out the experimental electronics and rendered the plane unusable for future missions. NASA was not pleased. Buffeted but invigorated by the successful but totally unauthorized flights, Joanne told the press that she felt fortunate to have seen meteorology develop from the "horse-and-buggy era" to the space age.[1]

Joanne Simpson (1923–2010) flew into severe weather systems, constructed computer cloud models, experimented on clouds and hurricanes, held two university professorships, mentored a generation of meteorologists, and supervised satellite remote sensing of the tropics. Her career spanned sixty-two years of major changes in meteorological science, from her role in training aviation weather cadets during World War II to the emergence of atmospheric science as an interdisciplinary, space-age field. She first focused her attention on the tropical atmosphere

---

[1] "Into the Eye of a Cyclone," 6; "First Lady of Tropical Meteorology."

First Woman: Joanne Simpson and the Tropical Atmosphere. James Rodger Fleming,
Oxford University Press (2020). © James R. Fleming.
DOI: 10.1093/oso/9780198862734.001.0001

in 1947, when little was known about its structure or dynamics. She utilized transformative new technologies: weather radar, digital computing, chemical cloud seeding, and satellite remote sensing to gather the information she needed. She dedicated her career to understanding the atmosphere better by modeling, measuring, and experimenting on tropical clouds and weather systems in regional and global field campaigns.

Most works in the history of science are not about meteorology, and most histories of meteorology focus on polar and temperate regions. Without exception, histories of meteorology examine the accomplishments of a cadre of male scientists. They also tend to marginalize or erase completely the contributions of female researchers. Joanne's work on the tropical atmosphere did not fit this pattern. Although the technical literature on tropical meteorology is immense, its history has not yet been written. Sources exist in thousands of scientific reports by hundreds of meteorologists. They include descriptive studies of monsoons, tropical and extra-tropical oceanic cyclones, and large-scale wind circulations. This study of Simpson follows the trail she blazed through this field from World War II into the first decade of the twenty-first century—a valuable narrative thread that is both unique and comprehensive.

The tropical atmosphere is filled with clouds, most of which look like small puffy cotton balls. These are called the trade cumuli, and they never grow large because they are being held down by a stable layer of air above them—the so-called "trade wind inversion," in which the temperature increases with height. Some clouds remain buoyant throughout their lifetimes and become towering monsters reaching into the stratosphere. These "hot tower" thunderclouds bubble up and then seemingly boil as they grow to their uppermost limit, generating heavy rains, strong winds, and immense ice clouds at the very top. Joanne's famous hot tower hypothesis, developed jointly with Herbert Riehl, alludes to the competition between a cloud's buoyancy and larger-scale environmental forces pushing it down. The hot tower theory valorizes those clouds that successfully break through the trade wind inversion to soar to the top of the lower atmosphere. They carry heat and moisture from the ocean's surface to the edge of the stratosphere and in doing so, generate lightning, thunder, intense rainfall, and strong updrafts and downdrafts. Joanne's work showed that towering thunderclouds are influential at all scales. They are incredibly important. They provide intense, episodic rainfall in the tropics, act as the spark plugs that drive hurricanes, influence trade wind patterns, and ultimately, link the wind circulations of the entire planet. The *metaphor* of hot towers points to the upward trajectory of Joanne's career. If buoyant air parcels are to grow into towering clouds, they must break through their environment and avoid dry air entrainment that would stunt their growth. This symbolizes the battles Joanne waged between her sky-high aspirations and the dark medical, psychological, and institutional forces attempting to drag her down. Joanne prevailed, reaching the pinnacle of personal and professional accomplishment and conditioning the atmosphere for further breakthroughs for women in meteorology and further developments in the understanding of the tropics.

Is Joanne Simpson best remembered as a pioneer woman scientist or the best tropical scientist of her generation? She was both, with the emphasis on best scientist. In the face of severe and constrictive social pressures, she earned a PhD in meteorology, the first woman in the United States to do so. Her obituary in the *Washington Post* erroneously referred to her as the "first female meteorologist to earn a doctorate," an error often repeated in print and even, in her old age, by Joanne herself.[2] The first woman to receive a doctorate in meteorology was Marie Dietsch (*b*.1890), who studied in Germany under Vilhelm Bjerknes at the University of Leipzig Geophysical Institute and received her D. Phil. in 1918.[3] The following table lists other women with advanced training in meteorology.

During World War II, approximately fifty women received nine months of training in the so-called "A" Course, given at several US universities to train aviation cadets to become military weather officers. As instructors in the synoptic laboratory, their main duties involved plotting and analyzing weather maps by hand, which were used to teach forecasting. After the war, the women were supposed to go back home, get behind the mop, and have babies, which nearly all did. John Lewis concluded, based on his thorough survey, that less than 10% of the 200 US women who were trained to be forecasters during World War II

---

[2] Sullivan, "Joanne Malkus Simpson"; Atlas et al., "Obituaries: Joanne Simpson, 1923–2010; Houze, "Joanne Simpson (1923–2010)"; Hardaker, "Dr. Joanne Simpson"; Atlas and LeMone, "Joanne Simpson"; and "First Lady of Tropical Meteorology."

[3] Dietsch, "Untersuchungen über die Änderung des Windes"; a thesis in mathematics and geophysics investigating the change of the wind with height in cyclones.

Women with Advanced Training in Meteorology

Name, degree, field, date, institution (supervisor)

Marie Dietsch, D. Phil. mathematical geophysics, 1918, Leipzig (V. Bjerknes)[4]

Elen Elaine Austin, BSc. maths and natural sciences, 1918, Cambridge, (N. Shaw)[5]

Luise Charlotte Lammert, PhD climatology, 1919, Leipzig (V. Bjerknes)[6]

Anne Louise Beck, MS geography, 1922, Berkeley (V. Bjerknes)[7]

Katharina Dorffel, PhD bioclimatology, 1935, Leipzig, (L. Weickmann)[8]

Fern Kirkman, MS meteorology 1939, NYU

Margaret Whitcomb, MS meteorology 1940, MIT

Inger Marie Bruun, MS meteorology, 1942, Oslo[9]

Karen Gleim, MS meteorology 1942, MIT[10]

Pauline Austin, PhD physics 1942, MIT (J. Stratton)[11]

Joanne Malkus, PhD meteorology 1949, Chicago (H. Riehl)

Patricia Langwell, PhD meteorology 1950, NYU[12]

---

remained in meteorology in any capacity.[13] A handful of them, Joanne included, continued their education and earned advanced degrees in meteorology, but every possible obstacle was put in their way, ranging from refusal of scholarships to downright hostility within the academy.[14] Although the Servicemen's Readjustment Act of 1944, commonly

---

[4] Other German women who completed their studies before the mid-1930s include Christine Stellmacher (Münster), "Über den Einfluss," Gertraud Richter (Leipzig); K. DeBary, Sigrid Soder, and Ingrid Weiss (Frankfurt), and Erna Plendl (Graz); Lewis, "WAVES."

[5] Barrett, *Women at Imperial College.*

[6] Börngen, "Lammert, Luise Charlotte"; Lammert, *Der mittlere Zustand der Atmosphäre bei Südföhn.*

[7] Beck, "An Application of the Principles"; Fleming, *Inventing Atmospheric Science,* pp. 52–9.

[8] Lettau, H. [and Käte Dorffel Lettau]. "Interview," by Sharon Nicholson; Dorffel, "Die physikliche Arbeitsweise."

[9] Meteorologisk Institutt, Norway and Magnus Volsett, personal communication.

[10] Lewis, "WAVES."

[11] Hinkel, "Celebrating Pauline (Polly) Morrow Austin."

[12] Langwell, "Inhomogeneities of Turbulence"; Simpson Papers, 9.9. Work Notebook #4: Woods Hole, October 1953–March 1954.

[13] Lewis, "WAVES."

[14] Simpson, "Women in the Atmospheric Sciences"; Turner, "Teaching the Weather Cadet Generation."

known as the "GI Bill," provided educational opportunities to both women and men, colleges gave overwhelming preference to returning male soldiers. Joanne and her near contemporaries still had to break through sexist barriers, expend enormous amounts of energy, and persevere through difficult times to attain any recognition as academic women.

Joanne was the first woman to receive NSF grant support in atmospheric science and the first woman to win the Carl-Gustaf Rossby Research Medal, the top award given by the American Meteorological Society (AMS). In addition, she was the first female meteorologist inducted into the National Academy of Engineering, the first female president of the AMS, and the first woman to win the International Meteorological Organization (IMO) Prize. She did so by combining intellectual curiosity with hard work and dogged determination, all in a field dominated by men. Like Marie Curie, Joanne's accomplishments were acknowledged in her lifetime. Similar to Curie, who coined the term "radioactivity," Simpson identified "hot tower" convection as a fundamental phenomenon. However, Joanne did not suffer from the "Madame Curie complex" and did not feel pressured to embody sacrifice, motherhood, devotion, altruism, humanitarianism, or other supposedly feminine virtues.[15]

Given her many accomplishments, what is currently known about Joanne among historians of science, meteorologists, and the general public is not close to what should be known about her. Several short, formulaic biographical sketches offer bare-bones outlines of Joanne's scientific accomplishments, but omit the flesh-and-blood details.[16] The third volume of Margaret Rossiter's notable trilogy, *Women Scientists in America,* contains a brief recognition of Joanne Simpson: "among the few women scientists at NOAA for a time was the pioneering meteorologist Joanne Simpson, who moved around a lot in her career and also worked at NASA's Goddard Space Flight Center, in Greenbelt, Maryland." Sarah Dry's book, *Waters of the World,* aims higher by including a chapter on Simpson's life and work based, in part, on archival materials.[17] Most media accounts reinforce the overly narrow impression that she was an

---

[15] Bamzai, "NSF's Role"; Quinn, *Marie Curie*; Des Jardins, *Madame Curie Complex.*

[16] Tao et al., "Research of Dr. Joanne Simpson"; Atlas and LeMone, "Joanne Simpson"; Weier, "Joanne Simpson (1923–2010)."

[17] Rossiter, *Women Scientists in America: Forging a New World Since 1972,* p. 179; Dry, *Waters of the World,* pp. 147–88.

ambitious, single-minded scientist. Even her colleagues and close friends, who were aware of some of her personal challenges, generally perceived her as calm and collected, a pretty cool character. Her response: "Nothing could be farther from the truth."[18] Clearly a deeper dive is needed to situate this history.

In 1990 the Schlesinger Library of Radcliffe College honored Joanne as a pioneer woman in science and invited her to deposit her archival materials there. The Simpson collection is candid, comprehensive, and personal, yet it is not self-serving. Its claims can be verified by cross-checking with other sources, including other archives, published materials, and interviews. The Simpson Papers are remarkable for including Joanne's never-before-seen private diaries, revelations about her health, and her accounts of the conflicts within her family. They also contain captioned photographs, scrapbooks, research notebooks, and several artifacts, including a three-dimensional cloud model, a silver iodide flare, and her ballet shoes. She donated materials to the archive through 2005, adding descriptive and often self-reflexive accession notes to the collection, leaving a rich visual, textual, and material trail for the use of historians. In addition to Joanne's life, the collection features extensive materials on the tropical atmosphere and on women in meteorology. It has important implications for gender studies, mentorship, and for the use and construction of archives. She kept a record of her research thoughts, even those that led to a dead end or down an apparent blind alley. Some of the abandoned ideas eventually blossomed after new techniques and observations gave her totally different insights on the problem. This happened over and over again. Joanne thought her detailed and disciplined practice provided a useful lesson for young scientists and science historians.[19]

The potential of the Joanne Simpson Papers lies in their power to recast the use of a life in science to tell larger stories about science and society. She wrote to the archivist: "I have deliberately kept my personal life as private as possible, and hence, if I should die before I finish, the material I am starting to send you now on my personal life would be lost, as little of it is known by anyone else. To understand how a

[18] Simpson Papers, 1.14. Family history: Simpson's narrative re: difficult childhood, lifelong depression, detailed photograph captions with commentary, January 1996.
[19] Simpson Papers, 9.2. Notebooks, Research, Studies and Ideas of Joanne Simpson, 1990–1993.

woman, or a man, for that matter, creates original work in any field, it is necessary to penetrate the emotional masks, and my masks have intentionally been hard to penetrate." Joanne proposed to take them off in her archival donations, "in so far as that is possible."[20]

Additional archival documentation comes from two long oral histories (by Margaret LeMone and Kristine Harper), as well as the archives of the University of Chicago and the Woods Hole Oceanographic Institution. The author and his assistants also conducted audio interviews and produced short video clips posted on a YouTube playlist: *Joanne Simpson Vignettes*, that feature many of her collaborators. These interactions provide additional insights into Joanne's life, work, and especially, mentorship.

This book is about Joanne, her career prospects as a woman in science, and her relationship to the tropical atmosphere. This braided narrative has three strands: Joanne's life and work in science, the shifting fortunes of women scientists in the twentieth century, and a first-ever excursion into the history of tropical meteorology. These multifaceted and interacting textual streams form a complex dynamic system—a system that frequently displays surprising emergent properties through the interaction of her ascending career trajectory, often turbulent personal and structural constraints she confronted, and new scientific opportunities opening up in the tropics.

Joanne's life spanned almost eighty-seven years of nearly constant turmoil, initiated by childhood neglect and abuse from her mother, battles with sexism in the workplace, a constant pursuance of love and acceptance, three marriages, a decades-long extramarital affair with a colleague, recurrent debilitating migraine headaches, and frequent bouts of depression. In her 30s, Joanne was passionately in love, for time and eternity, with Claude Ronne, her photographer colleague at Woods Hole. For the final four-and-one-half decades of her life, she found abiding love and professional fulfillment with her third husband, hurricane specialist Bob Simpson. This book, while anchoring Joanne's story in her intellectual accomplishments, provides a fuller picture of her personal and professional life, all in order to understand better her experiences and accomplishments as a highly decorated scientist.

Joanne Simpson was born Joanne Gerould and changed her name after each of her marriages: Joanne Gerould Starr, Joanne Starr

---

[20] Simpson Papers, 1.14.

Malkus, and finally, Joanne Simpson. She was the daughter of a first-wave feminist, Virginia Gerould, but did not participate in subsequent "waves." Instead, in the rising tide of feminism, she made big waves of her own, for professional women in general, but primarily for women in meteorology. Joanne advised women to work twice as hard to get ahead in the male-dominated world of science. Joanne battled against institutionalized sexism by focusing on the task at hand, not choosing to risk facing additional discrimination as she confronted the barriers standing between her and her career in meteorology. At the start of her career, her decision to change her last name would have increased her visibility within the meteorology community by associating her with a known male meteorologist. Her first husband, Victor Starr, was recognized as a rising talent at the University of Chicago. When she married physicist Willem Malkus, she kept Starr as her middle name, not to honor her former husband, but to provide a sense of continuity in her career. In 1965, with her reputation as Joanne Starr Malkus firmly established, and following a bitter second divorce, she took the name Simpson. Secondary sources occasionally refer to her as Joanne Malkus Simpson, a name she never used. She published under all three married names, however, and her gravestone in Rock Creek Park Cemetery, Washington, DC, displays the name Joanne Gerould Simpson, a final return to family roots. In this book she is called, whenever possible, "Joanne."

Joanne was raised in Cambridge, Massachusetts, where she benefitted from educational opportunities but suffered emotional neglect, especially from her mother. At age 17 she enrolled in the University of Chicago and took a number of introductory courses, including astrophysics and psychology. In 1942, at the end of her sophomore year, she met Carl-Gustaf Rossby and joined the war effort, working to train aviation cadets for weather forecasting. She pursued advanced training in meteorology, but the all-male Chicago faculty opposed her at every step. Her interest in the tropics was piqued by a class she took with Herbert Riehl, and the two began a long-term and fruitful collaboration.

With her PhD in hand, and seeking first-hand experiences flying in clouds and the tropical atmosphere, Joanne moved to Woods Hole, where she worked for ten years. In collaboration with Riehl, she developed her signature "hot tower" hypothesis of convection. Her rapid rise in the profession led to an offer in 1960 of a full professorship of meteorology at UCLA, but emotional turmoil left her life in shambles and resulted in her leaving the university after only three years. Joanne settled down with her third husband, hurricane expert Robert Simpson, and accepted a position directing the experimental meteorology laboratory in the National Weather Service. Ten years later, Joanne again flirted with the offer of a named chair at the University of Virginia, but the position was not right for her or her husband. She finally found contentment and professional fulfillment at NASA, where she supervised the development and launch of the Tropical Rainfall Measuring Mission satellite and served as a mentor to many, many colleagues.

Joanne sought equal opportunities for women in science, critiqued those who would oppress her, and broke through professional and societal barriers to develop a new understanding of the tropics. There is an immense and growing literature on the subject of women in science, but none on women in meteorology. Of the handful of accomplished women trained in meteorology in the early decades of the twentieth century, each had been exceptional, and all in a male-dominated field. All had to confront the metaphorical limit referred to as the trade wind inversion—the glass ceiling of meteorology. Joanne's science showed how clouds could break through the inversion and reach new heights. Joanne too broke through: in her life, in her science, and for those who followed her.

1

Dark Clouds at Dawn

"You have to be lovable to be loved."

Virginia Gerould

Winter's lingering ice and snow were in rapid retreat in Boston on Friday, March 23, 1923, the day Joanne came into the world, two weeks premature. Spring showers greeted the new baby, but darker clouds were on the horizon. Rather than being elated at the arrival of her first child, Joanne's mother, Virginia, felt emotionally and physically drained and quite resentful. Her pregnancy was unexpected and unwanted, coming so soon after her marriage to Russell Gerould, a graduate of Harvard University and a rising aviation reporter for the *Boston Herald*. Virginia Isabella Vaughan was born in 1898 to a prominent Watertown family. She spoke of an unhappy childhood, suffering from frequent and severe migraine headaches, experiencing lack of affection from her own mother, and receiving harsh and sometimes public criticism from her father.[1] Like her mother, Virginia attended Radcliffe College, graduating in 1919 with a degree in English before landing her first job with the *Boston Globe*. Virginia was highly motivated and interested in pursuing a career in journalism or perhaps public relations, but the arrival of Joanne forced her to give up a promising writing career, which she was never able to resume in the way that she wanted. Harboring deep resentment and wanting nothing to do with Joanne, Virginia reluctantly nursed the child for nine months. Believing that nursing prevented conception, Virginia accidentally became pregnant again when Joanne was about three months old. This pregnancy was terminated by abortion, which, at the time, was illegal. By no fault of her own, Joanne and her mother had started out on the wrong foot. Her mother told her, at a very tender age, that she was a very ugly baby, puny, and "covered with fur"[2] What a horrible start in life!

[1] Simpson Papers, 1.11. Family history: Simpson's narrative re: mother Virginia Vaughan Gerould's family, March 1991.

[2] Ibid.

First Woman: Joanne Simpson and the Tropical Atmosphere. James Rodger Fleming, Oxford University Press (2020). © James Rodger Fleming.
DOI: 10.1093/oso/9780198862734.001.0001

Joanne's memories from her early years of life included lying flat in a pram, her mother pushing her rapidly away from herself shouting, "Whee!" Joanne also recalled seeing her parents demonstrate affection for each other, on at least one occasion, hugging in the kitchen. She could not remember feeling any warmth or affection from her mother; however, she clearly recalled experiencing her mother's furious anger. Virginia, who did not work outside the home, hired a house manager, Mrs. Forrest, whom Joanne called "Foffie," to cook, clean, babysit, and serve as a surrogate mother. Joanne was just too young to miss the much greater affection she learned later that most mothers bestow upon their children.[3] Joanne revealed the lifelong tension in her relationship with her mother in the caption to this 1924 photograph she composed in 1990 (Figure 1.1).

Figure 1.1 "July 1924 at Wellfleet already a 'boat person' at age one with Mother, age 26. A model boat is distracting me. My mother disliked boats of the rowing and sailing variety. Concerning the luxury cruise liners, it was the opposite story." Simpson Papers, 455, Photos. PD.2-22.

---

[3] Simpson Papers, 1.13. Family history: Simpson's narrative re: overview of childhood from six months to eight years old (1920s–1930s), September 1994.

Despite the neglect, as a young child Joanne recalled happy times. She adored her father, then in his early career, and relished his companionship. He was proficient at math, an accomplished musician, an excellent tennis player, and an outdoor person, "interested in nature, the oceans, atmosphere, hiking, and things like that."[4] He took Joanne swimming and rowing, and sometimes let her sit on his lap and steer the car. He was affectionate, often kissing Joanne good night. Russell Gerould's career was quite successful; he created the role of aviation editor of the *Boston Herald* and appointed himself to fill it. He flew with many of the early airmail pilots and had a whole scrapbook of pictures of early aviation. One photograph showed Russell and Virginia boarding an American Airlines Ford Trimotor, the first scheduled flight from Boston to New York. They both sported raccoon coats; Virginia wore a close-fitting bell-shaped "cloche" hat—visible symbols of their privileged cultural status. At age four, Joanne remembered her father in his underwear joyfully dancing around the apartment screaming "He made it! He made it!" His personal acquaintance, Charles A. Lindbergh, had flown solo across the Atlantic. Lindy was not only a great hero to him but also an acquaintance. Russell flew as a reporter on the next to last flight of the ill-fated dirigible Akron, which crashed in high winds off the New Jersey coast in 1933. He took Joanne flying at age six in a two-passenger Curtiss Robin from what was then called Jeffrey Field. Joanne treasured a photograph of her (at age two) sitting on her father's lap (Figure 1.2).

Both the Vaughan and Gerould families were solidly upper-middle class with roots in colonial America. Walter Vaughan (called "Gubby" by his granddaughter Joanne) was the president and senior partner of the W. C. Vaughan Company, a wrought iron manufacturer with headquarters in Haymarket Square, Boston. His wife Mosetta Stafford Vaughan, or Etta (called "Gubaga" by Joanne), was a pillar of the Watertown Unitarian Church and active in the Historical Society. The Vaughans employed two servants, drove a fine car, summered on the coast, and regaled Joanne with accounts of their extensive travels in Europe.

Joanne's paternal grandmother, Florence Russell Gerould, was born in Nantucket of French Huguenot stock, the daughter of one of the last sailing captains and boat owners to make a living by whaling. She never went to college but was highly erudite, a well-read and well-travelled person. She was deeply religious, a pillar of the First Unitarian Church

---

[4] Simpson, "Interview," by LeMone.

Figure 1.2 "IMPORTANT picture July 1925 with my brilliant reporter (*Boston Herald*) father Russell Gerould, not quite 25 years old." Simpson Papers, 455, Photos. PD.2-27.

in Harvard Square. Grandfather Charles Gerould, who died rather young, also came from a New England whaling family. He was a "shadowy figure" to Joanne and was committed to a mental institution by the time she was five.[5] Joanne was fascinated by the nineteenth-century painting of a whaling ship by the noted artist Benjamin Russell that had been passed down from her whaling ancestors. Her father had always told her that it would be hers someday, but this was not to be.

Extended family was important to Joanne. For her first seven years she lived with her parents in a third-floor apartment at 76 Oxford Street, Cambridge, in a building owned by Grandmother Gerould, who lived across the street and often invited Joanne to sleep at her house. Her father's brother and his wife, uncle Dick and aunt Eleanor, both of whom she liked immensely, lived just below on the second floor.

[5] Simpson Papers, 1.11. Family history: Simpson's narrative re: mother Virginia Vaughan Gerould's family, March 1991.

Joanne's parents owned expensive furniture, fine china, crystal goblets, linens, and other accompaniments of gracious living. They ate their daily meals on blue and white china plates depicting star-crossed lovers on a bridge. Joanne was close with her grandmothers. They were major positive influences on her life, instilling in her the virtues of grit, resilience, perseverance, and hard work. Grandmother Gerould took Joanne to church many Sundays and once gave her a 5-dollar gold coin as an incentive to read the entire Bible, the King James Version; she just loved it. Much later, in school, Joanne took a year-long course on the Bible as literature. Although both of her parents were agnostics, Joanne turned out, if not overtly religious, then certainly not anti-religious. She was accepting of a wide range of personal religious sentiments, but opposed censorship and fanaticism in any form.[6]

Joanne attended kindergarten and first grade at the Lincoln-Field school in what had been a large mansion on Avon Hill, Cambridge. In Joanne's words, "The proprietors, Miss Lincoln and Miss Field, were two dried up old maids who regarded me as a problem child and told my mother that I really did not belong there. I cannot remember at all what I did that was so bad or in what way I was a problem, although I am sure that I was."[7] Joanne had taught herself to read at about age four, and was, from then on, an avid reader, for both pleasure and escape. She remembered and appreciated one good aspect of her mother: that she took her to the public library every few weeks to stock up on books. At about this time, Gubaga realized that Joanne had poor distance vision when she could not read the destination sign on an approaching streetcar. She was diagnosed with acute myopia and had to wear glasses. She blamed her mother for buying the cheapest and ugliest frames possible and forcing her to carry the glasses around in an awkward gray case suspended on a belt under her clothes. The other children teased her about the glasses, calling her "four eyes." Yet she was amazed to find that trees had leaves that she could actually see; she had been walking around in a blurred world.[8] Her mother controlled

---

[6] Simpson Papers, 1.13. Family history: Simpson's narrative re: overview of childhood from six months to eight years old (1920s–1930s), September 1994; R. H. Simpson, *Hurricane Pioneer*, 126–7.

[7] Simpson Papers, 1.14. Family history: Simpson's narrative re: difficult childhood, lifelong depression, detailed photograph captions with commentary, January 1996.

[8] Simpson Papers, 1.13. Family history: Simpson's narrative re: overview of childhood from six months to eight years old (1920s–1930s), September 1994.

everything, with no backtalk allowed, including Joanne's short-cropped hairstyle, clothing, and most everything else. Nearly all her girlfriends had long hair in braids, but Joanne's mother would not allow it, saying it would be too much trouble for her.

Joanne's parents were not happy. Soon after their marriage, Russell began drinking heavily, in the depths of prohibition, frequenting illegal speakeasies with his journalist colleagues. Virginia and her family found this behavior particularly offensive. In a futile attempt to save their marriage, Virginia decided to have a second child. Daniel Charles Gerould, nicknamed Jerry, arrived on March 28, 1928, and Joanne's life began a very painful downturn which she attributed to the birth of her brother. Jerry soon contracted croup and then whooping cough, and Virginia developed a breast abscess. Joanne contracted measles around the same time and was forced to live at Gubby and Gubaga's house, under the care of "a very cross and stuck-up trained nurse, Miss Duffy."[9] Joanne recalled being feverish and calling out in tears for her father and mother, but never seeing or speaking to them for weeks at a time. Virginia was in the process of weaning Daniel, but Joanne did not know what "wean" meant. She thought it might be something fatal and did not dare ask.

The emotional turmoil did not subside when Joanne returned home. Virginia was a very neurotic, needy person as a result of her loveless upbringing, poor health, and tumultuous marriage. She was very tense, even snippy with Joanne, and would not let her help take care of Jerry. Instead, Joanne watched her mother dote on the baby—nursing, cuddling, and bathing the chubby red-haired boy she wanted so much. Her mother made it abundantly clear to Joanne that she was the less-favored child. Joanne vividly recalled the most traumatic event of her young life when, in 1929, her mother was giving Jerry a bath, snuggling him and kissing him in the process. According to Joanne, "She had never snuggled and rarely kissed me. I asked, 'Why do you love him so much and don't love me?' Her reply seared like a heavy corroding acid in my chest for decades after: 'I love him because he is lovable. You have to be lovable to be loved'. The way she spoke gave me the clear message that I was unlovable as a permanent condition. I spent a substantial part of my life trying to make myself lovable or even likeable to her."[10]

[9] Ibid.
[10] Ibid.

Figure 1.3 "A sad looking girl who indeed felt sad at the time. I was a dejected child when my brother was born, and I was very sick with the measles. Photo taken [spring 1928] opposite 76 Oxford Street near the Agassiz School with mother's [teddy] bear, which she only allowed me to hold on special situations, such as being sick." Simpson Papers, 455, Photos. PD.2-58.

This was a devastating emotional blow. Jerry's many illnesses in his first two years resulted in Joanne's living with her grandparents for weeks at a time. Her insecurity about being unlovable would not be quickly resolved. Joanne even kept a ghostly photograph of her mother's empty bedroom and chaise lounge where she nursed and cuddled Jerry, adding to the caption, "And sibling jealousy took hold."[11] Joanne had become a sad and neglected child (Figure 1.3).

Happy Times at Humarock

There were, though, happier times that instilled in Joanne a love of nature. For her first eight summers, the family rented the same cottage

[11] Simpson Papers, 455, Photos. PD.2–59.

in Humarock, on a marshy estuary near Mansfield on the South Shore. Joanne referred to these as the "wonderful summers," and labeled her photographs from the period, "the magic days of childhood."[12] (Figure 1.4). The unfinished cottage, with a screened porch and garage, was located at the end of a bumpy dirt road, partly surrounded by woods. It had indoor plumbing and a two-burner kerosene stove, but no refrigerator. There was no electricity the first summer, and no telephone until the summer of 1929.

Joanne's bedroom was located on the west side of the cottage. It had a sloping ceiling and a view over the salt marsh, with a south-facing

Figure 1.4 "Rabbit appears as inducement to sit still for my father to photograph us at Humarock. This is the best childhood picture of me. There was a hassle about taking this picture (I think I was dragged there because a picture of my baby brother was wanted). I know I was putting up resistance or I would not have the rabbit, which was one of my mother's favorite toys from her childhood, which I only got to hold on special situations such as being sick." Simpson Papers, 455, Photos. PD-2.

[12] Ibid.

gable over the estuary. Joanne loved to watch the boats go by, even when the noise of motorized boats interrupted her afternoon naps. She called the nearby woods "the wild west." It was a place to explore, look for blueberries, or curl up with her favorite book. She especially enjoyed Robert Louis Stephenson's *Treasure Island*, the Dr. Doolittle talking animal series by Hugh Lofting, and Andrew Lang's books, full of fairies, witches, ogres, and other magical creatures. Joanne carried the sights, sounds, and smells of that forest and the salt marsh with her throughout her life.

Joanne loved the water. Her father encouraged her to swim when she was aged about three by throwing her off his shoulders, and by age six she had become a strong swimmer. She had no fear of the water, even when it might have been warranted. Joanne often played in the nearby estuary where she could observe the behavior of snails, clams, ducks, geese, and marsh birds at all stages of the tide. One of her favorite childhood activities was rowing. A small rowboat went with the cottage, and it was a rare day that she was not using it in some way. She also remembered her first exhilarating experience sailing, apparently in a fairly strong wind. In all of these adventures, Joanne's mother paid little attention to how long Joanne was gone or where she had been. Virginia was not a big fan of the cottage and much preferred tea parties with friends; she did not like having to cook, found the noise of the motorboats annoying, and was not much interested in boating or swimming. Joanne remembers her making disparaging remarks about some of the neighbor children who summered with their families, referring to their little cottages built close together as "junior towns."

However different Joanne and her mother were, they both loved animals of all kinds. The family almost always had a cat or two, a dog, and various other pets. Her mother would go out of her way to see that hurt or starving animals received proper care. On a drive to Plymouth, Joanne vividly remembered the car ahead of them striking a dog. Virginia insisted the driver take the seriously injured dog to the nearest veterinarian, but he refused, saying he was in a hurry. Joanne and her mother tried to put the dog in their car, but it died. Joanne cried all the rest of the way and was not scolded for it. Much later in life, Joanne regarded caring for animals as her mother's most endearing trait and was grateful for at least one aspect of life they could share.

In 1929, Joanne and her summer friends held a "fair" on the big beach at Humarock for the benefit of the Angell Memorial Animal Hospital.

Most of the contests cost a penny, the highest-priced game being darts. At 5 cents the winners got small candies or small toy animals. Other games included Beano (now called Bingo) and shooting marbles in a circle. The children made almost $5—more than they expected and now worth more than $70. When they presented the proceeds to the animal hospital, one of the head veterinarians thanked everyone and wrote a wonderful letter. Joanne remembered this day and cited it as one of the most joyful in her early life.

In the summer of 1931, Joanne suffered an attack of appendicitis. She was writhing in pain on the dirt floor of the garage at Humarock and implored her three-year old brother to go fetch their mother. Joanne recalled her mother issuing a stern order to her to "get up and get in the house." When her condition worsened, her mother telephoned their doctor in Cambridge and brought her to the Mount Auburn Street Hospital the next day. On the drive over, she remembered lying on the floor of the old Dodge feeling very ill. The doctor met them at the hospital and signed Joanne in. Her mother thanked him, turned, and left with her brother. Joanne assumed she said goodbye, but did not recall that her mother gave her a hug or kiss, or seemed in any way perturbed by her illness, except that it was a nuisance. To add insult to injury, her mother did not visit or telephone Joanne during the one-week hospital stay. Her one comfort was the loan of a Nancy Drew mystery novel, her first, which provided a pleasant distraction.[13]

Virginia hired Ita McKenna, age 17, as a full-time housekeeper, cook, baby sitter, and eventually, complete home manager. She and Joanne took an instant and intense dislike to each other. Within a few years, Ita controlled the children's world. She clearly favored Jerry and reported on Joanne's bad behavior to her mother, making her miserable home life even more miserable. Joanne compensated, in part, by keeping her distance from home and investing heavily in her schooling. From 1930 to 1940, with the exception of a year in Washington, DC, Joanne attended the Buckingham School in Cambridge—a private, all-girls school where an intelligent and somewhat rebellious child could flourish. She walked the two miles to school early in the morning and immersed herself in her studies and extracurricular activities all day, often staying until after 5 o'clock. Her teachers made learning a fascinating adventure

---

[13] Simpson Papers, 455, Photos. Most memorable experiences of my first 8 years which do not have pictures to illustrate them.

and opened doors for her future successes. She called it "the best thing that ever happened to me."[14] Her teachers were more than imparters of knowledge; they were forgiving parent figures who responded to her often rebellious behavior with a combination of strictness and love, both of which she desperately needed. Joanne remembers behaving badly many times, once walking a 20-foot high tight-wire strung in a campus tree. Her teacher, Miss Valiant, gave her a severe dressing down, but did not demean or frighten her. She inspired Joanne to behave better in order not to worry others.

Washington, DC

In 1931, Joanne's father obtained the plum job of being the Washington Correspondent for the *Boston Herald*, and the whole family, including Ita, moved to Washington, DC. They rented a two-story Tudor-style home in the Columbia Heights area, near the terminus of the Mount Pleasant streetcar. Joanne was homesick and longed for the Buckingham School, the familiar pathways of Cambridge, and the cobblestone streets of old Boston. She was enrolled as a third-grader in the Bancroft School, a completely segregated public school. She (as well as much of the country) had not yet realized that race was a serious issue, but later became a staunch supporter of the civil rights movement. Her new classmates were fairly hostile to her at first, but she outperformed them and, in rapid succession, was quickly advanced to high third grade, fourth grade, and high fourth grade, changing teachers and classmates with each move. The frequent moves were no doubt unsettling. Nevertheless, at this point, fearing that it might be harmful to Joanne to be in a class with significantly older children, her mother decided to take her out of public school and enrolled her in fourth grade at the Potomac School, an exclusive private academy located in the DuPont Circle area. Unfortunately, here too she was met with hostility from the other girls, who teased her unmercifully, especially about wearing glasses and speaking with a "Yankee" accent.

Joanne recalled being miserable at school and at home. She became rebellious. Ita and her little brother reported every misdeed of hers, however minor. On one occasion, Virginia responded by taking away

---

[14] Simpson Papers, 1.13. Family history: Simpson's narrative re: overview of childhood from six months to eight years old (1920s–1930s), September 1994.

her one and only doll, Mary Lee. Joanne reacted by running away from home, at least for the day. She had enough money to buy cupcakes at the drugstore and ride on the streetcars, whose routes she had memorized. She would venture into what were called then "colored" neighborhoods and talk to elderly people sitting on their porches. Sometimes she spent the day with playmates in nearby Rock Creek, Park, creating dams and whirlpools and visiting the zoo. She usually came back home very late at night and went straight to bed, with nothing more said about it until the next time. Joanne recalled running away five or six times. Once, when she threatened to run away permanently, her mother offered to pack her suitcase, but with nowhere to run to and school work to do, Joanne just left for the day and came home again many hours later.

Virginia enjoyed an active social life as the wife of the Washington representative of an important out-of-town newspaper. The Geroulds were regulars on the cocktail circuit and met all kinds of celebrities, including Pierre Laval, Prime Minister of France, who was the 1931 *Time* magazine man of the year. Joanne also benefitted from her father's status. She recalled meeting Massachusetts Representative Edith Nourse Rogers and sitting at her desk in the House chamber. Joanne was embarrassed and insulted, however, when her mother repeatedly told an apocryphal story about this visit to Congress. She claimed that someone had asked Joanne the difference between Republicans and Democrats, and Joanne supposedly said, "The Republicans have fatter stomachs and the Democrats argue more." Although this seems funny now, as an eight-year-old she was infuriated at this insult to her intelligence. When she contradicted her mother, it only made Virginia angry. Later she learned, most of the time, to bite her tongue and distance herself from the situation as soon as possible.

Back to Cambridge

As school was finishing in 1932, Joanne learned that her father had been promoted to city editor, a big accomplishment, and that the family would be moving back to Massachusetts. She could hardly restrain her joy. Everyone was dancing around the house and hugging each other. The next month, when the old family car finally crossed the state line of Massachusetts, Joanne insisted on getting out to kiss the ground and hug the nearby trees. She then realized how homesick she had been.

Virginia still had creative aspirations and was ready to go back to work. She was heavily influenced by Margaret Sanger and supported a woman's right to birth control. Joanne's unwanted birth represented Sanger's main contention that "every child should be a wanted child." Virginia's commitment to feminism led to her employment as head of public relations for the American Birth Control League. Her new job posed a problem for the family. At the time, any kind of contraception or information regarding it was illegal in Massachusetts. Russell had accepted a new position as head of public relations for Governor Leverett Saltonstall, and half of the residents of Massachusetts were Catholic. Moreover, the Bishops reviled Planned Parenthood and strongly denounced Margaret Sanger and her movement. This forced Virginia to work under her maiden name, "Miss Vaughan," in order to avoid embarrassing her husband.[15]

In May 1937, Joanne's mother announced that the family would be staying in Cambridge for the summer. She claimed the reason was lack of money, but given that they were by no means poor, Joanne suspected and later found out that the real reason was another man. Joanne was quite disappointed not to be living near the beach for the summer, as they did in previous years. At age 14 she knew, or at least intuited, that lack of structure in her life led to depression, which she was just beginning to experience. So she went looking for a summer job at a vacation place. She wrote to a number of national parks, but they employed no-one under 16. When Joanne raised the possibility of waiting on or clearing tables at a restaurant on the south or north shore of Boston, her mother pointed out that she probably could not earn carfare and lunch money, much less room and board, since she had no marketable skills except occasional babysitting.

Following several rejections, Joanne landed her first summer job working for her Cambridge neighbors, the Coit Family, at their primitive woodland camp more than an hour's drive away, near Rockport on Cape Ann. The camp had running water, but no electricity. Joanne took care of the children—Mary, Kaki, and Sandy, aged 8, 6, and 2. She regularly cooked breakfast for the whole family, helped with the other meals, and cleaned up the camp for a salary of $3 per week. She was free on Thursday and Sunday afternoons and evenings, and was allowed an

---

[15] Simpson Papers, 1.11. Family history: Simpson's narrative re: mother Virginia Vaughan Gerould's family, March 1991.

hour each day to herself. The routine was tiring. Cooking and house-work were not too bad, except for the 6:30 am breakfast, where Mr. Coit had to have one perfectly poached egg with perfect toast before he caught the morning train to Boston. That had not sounded difficult when Joanne had discussed it in Cambridge, but she had never tried to cook anything on a kerosene stove. The eggs finally began to come out right, and Mr. Coit tactfully accepted the toast, even if it was sometimes scruffy.[16]

All three children were bright, active, and inquisitive. Joanne enjoyed taking care of them, but wondered many times if she had the maturity to supervise them properly. Mary and Kaki asked questions about virtually everything from stars to how babies were made. Joanne remembered swimming off the beach with the two girls in water over their heads and getting back safely without their mother finding out. She also recalled how kind their mother was to let her blunder around in Gloucester Harbor with the girls in a sailboat in a light wind. Joanne read during her free hour while Sandy took her nap, but found it almost impossible to read in the evenings by the dim light of a kerosene lantern while surrounded by a swarm of insects. Usually she was too tired to do anything except flop into her cot behind the kitchen after the older children were settled for the night. Everyone bathed while swimming, followed by a quick rinse in a cold outdoor shower. Everyone went bare foot, and no one minded being a bit grubby.[17]

Joanne hoped to spend her free time with vacationing young people she had met in previous summers. Lacking a telephone or access to transportation, she had to walk the five miles into Rockport, which she did only occasionally. Once she watched artists painting on the beach and dove off the town pier. On other occasions she went to the public library to read. Once, a young man invited her to a movie, but he was unable to find the camp road or perhaps changed his mind. Virginia came to visit once that summer and took Joanne to a nice restaurant overlooking the harbor. Joanne had eagerly looked forward to the visit and saved up to buy her mother a gift, a beautiful painted wooden bowl from a local craft shop, but her feelings were hurt when her mother accepted it dismissively. Joanne never saw the bowl again. Toward the end of August, Joanne caught a bad cold, which developed into bronchial

---

[16] Simpson Papers, 1.5. Family history: Simpson's narrative: "My Mother and I," January 1998.
[17] Ibid.

pneumonia. Consequently, she had to go home for a week of proper care and bed rest. Soon, however, her mother insisted that she go back to the camp. In her weakened state, Joanne recalled, she was running on adrenaline, but did not wish to fail at her first job. Later that fall, Virginia accidentally left her diary by the telephone, open to the date of her Rockport visit. When Joanne glimpsed her name, she guiltily read the sentence in which it appeared: "saw Joanne today. Poor thing looked so dirty and disheveled—needs a clean apron." The rest of the entry went on about the man she loved, which Joanne did not read, not because of any high moral attitude, but because she was too distressed that all her mother seemed to be concerned about was her dirty apron.[18]

The start of the new school year in mid-September was a welcome relief. At the Buckingham School, Joanne delved into history, literature, and public affairs, but was especially fond of mathematics for its abstract intellectual challenges. Physical science attracted her less, since it seemed to be mainly about machines. She was involved in sports such as hockey (captain), basketball (captain), and archery, and was very active in the dramatic club (Figure 1.5).

Joanne credited the school with developing her intellectual curiosity, good study habits, and organizational skills. Buckingham provided

Figure 1.5 Joanne Gerould's high-school yearbook senior photograph and list of activities. Simpson Papers, 4.8. Simpson Symposium, February 9–13, 2003.

[18] Ibid.

life and career resources for its students, including a seminar Joanne attended on "Careers for Women." Her teachers required her to write a substantial paper virtually every week and speak before the entire school twice a year. She also edited the school's yearbook, the "Packet." This extract, written in her senior year at age 16, captures the essence of her adventurous spirit, authorial voice, and sensitivity to the power of nature:

> Monday, November 6. Yesterday it rained. Not just a slow, provoking drizzle, nor a steady purposeful rain, but the exhilarating kind of north-easterly, drenching, driving, stormy rain that makes you want to be on a boat or at least near the ocean. In the city such a storm is out of place. It drenches you right through the sissy rain-coats people wear in the city, and it tears silly silk umbrellas right out of their silly hands, and they say, "Oh dear, how I hate this AWFUL rain!" Ordinary rain depresses me, too, but when I am outdoors in a northeast storm, I feel excited and possessed by a strange sea fever. It reminds me of the time that we were ship-wrecked off Manomet. I remember how the wind kept getting stronger and stronger, and the strange noise the water made as it rushed by, and how queer and white and foamy it looked as it stretched out behind us. I can smell, in my imagination, the yeasty, fishy, salty smell that comes on a Northeaster, and I can hear the roar that the waves make as the wind piles them onto the sand. In the midst of a storm they are irregular and choppy, and break almost as far out as the horizon. It is only after the storm has blown itself out, and when there is a ground swell, that they come ashore in huge, quivering rollers, which get higher and higher as they get nearer land, and finally curve over and break in an almost straight line along the beach. These rollers make a crashing, pounding sound, which is quite different from the continuous roar of the breakers at the height of the storm…During the night, the storm let itself out and this morning, when I was walking to school, there wasn't a single cloud in the sky, and the sun seemed all the brighter because of the rain yesterday. The puddles were all yellow and muddy and full of fallen leaves, and other leaves were plastered to the sidewalk. Each of them had a little halo of dampness around it. At the field in the afternoon, the ground was slippery and people kept falling down in the mud…[19]

Many decades later, in 1996, when Joanne received the Buckingham's Distinguished Alumna Award, she remarked that "this award is the one that I will cherish most for the rest of my life." To understand that,

---

[19] Simpson Papers 4.8. Simpson Symposium, February 9–13, 2003, citing J. Gerould, *The Packet of the Buckingham School*, June 1940.

she continued, "I have to tell you what the Buckingham School did for me. It can be summarized in one sentence: I would not have achieved anything at all without the Buckingham School. I owe the school so very much."[20] In her typical candid manner, Joanne expressed her gratitude to the teachers who preserved her emotional equilibrium during a very difficult period when her mother had rejected her and her parents were at war with each other.

By the late 1930s, the marriage between Virginia and Russell Gerould had deteriorated so far that they were essentially separated. Russell spent substantial periods of time away from home, and Virginia was fully occupied with her paramours. She moved into an apartment in 1940 after Joanne went away to college. Their divorce was finalized in 1945. Russell remarried; Virginia dated, but remained single. For the rest of her life Joanne felt the tension in her relationship with her mother; Virginia never ceased expressing her resentment and frustration with Joanne. Joanne tried repeatedly to reconcile with her mother, at least in her own mind, by listing the positive things Virginia had done for her. She even tried to gain her approval by winning awards, but all her efforts were for naught. In her old age, Virginia formally disinherited her daughter.

Joanne recalled, at age 14, standing outside the Harvard/MIT Cooperative Society store (or COOP) in Harvard Square and making the strongly held determination that "I'm going to get somewhere and be somebody."[21] Her schoolmates thought she might turn out to be a Senator, resembling her remote ancestor, Daniel Webster, or perhaps a writer or editor. Her family expected her to attend Radcliffe, as had her grandmother Gubaga, a member of Radcliffe's first class, and her mother Virginia. In spite of the abuse and neglect she experienced as a child, Joanne credited her mother with doing some important things for her, namely, providing her with an amazing education, introducing her to other influential people, and encouraging in her a feminist way of thought, which she clearly put to use throughout her life.

Joanne worked in the summers from 1939 to 1941 at Jeffrey Field (now Boston Logan Airport) as an assistant to Crocker Snow, Head of the Massachusetts Aeronautics Commission, who was instrumental in establishing commercial air travel in New England. There, after completing

[20] Simpson Papers 2.6. Awards: "Milestones and Awards, 1980s to 1990s."
[21] Simpson Papers, 9.2. Narrative by Simpson, January 1994, re: description of undergraduate and graduate college career.

a required course in meteorology, she earned her pilot's license. These experiences, combined with her father's positive influence and her passion for sailing, resulted in her lifelong love affair with clouds and the weather.

Joanne was clearly college-bound, but she was undecided about a major. She knew she wanted to excel at something, anything, and she wanted to accomplish this far from home. She was an avid reader and greatly admired the great books curriculum and the educational reforms being instituted by Robert Maynard Hutchins at the University of Chicago. She decided to start anew there. Her childhood had been like a passing storm. But who has ever heard of a 17-year storm? Perhaps a cloudless sky or a more temperate climate lay just over the western horizon.

2

Chicago

Your work on cumulus clouds is a fine field for a "little girl" to work on.

C.-G. ROSSBY

In early September 1940 Joanne bid goodbye to her family and boarded the train from Boston to Albany to connect to the westbound Lake Shore Limited. Her 24-hour journey to Chicago paralleled the route of the venerable Erie Canal across New York, then along Lake Erie, with a trek across Ohio and Indiana to her final destination. She held a ticket to new challenges, new adventures, and a new start in life based on her accomplishments and hard work. From LaSalle Street Station, it was a 45-minute bus ride along scenic Lakeshore Drive to Hyde Park and her new home at the University of Chicago. As the stately buildings and green spaces of the campus loomed before her, Joanne realized this was her first real solo flight—a one-way ticket to a new, improvised, and hopefully much improved life and identity. Sailors return to the dock and aviators to the aerodrome, but for Joanne, there was no going back.

The University of Chicago opened for instruction in 1892 as a modern research university, combining an English-style undergraduate college and a German-style graduate research institute. Architecturally, the buildings resembled those of the University of Oxford, complete with towers, spires, cloisters, and gargoyles, on a campus designed by the legendary landscape architect Frederick Law Olmsted. Its first president, William Rainey Harper, established the tradition of rigorous academic training for people of all backgrounds and equal opportunities for both sexes. He established a community of inquiry composed of highly accomplished teacher-scholars and students who were granted complete freedom of speech and inquiry on all subjects. The university coat of arms depicts a phoenix emerging from the flames over the motto *Crescat scientia; vita excolatur* ('Let knowledge grow from more to more; and so be human life enriched'). Soon, Chicago's reputation

First Woman: Joanne Simpson and the Tropical Atmosphere. James Rodger Fleming, Oxford University Press (2020). © James Rodger Fleming.
DOI: 10.1093/oso/9780198862734.001.0001

soared as a place for world-class research and serious, hard-working, and very smart students. During the presidency of Robert Maynard Hutchins, in the 1940s the general education curriculum shifted from large lecture format to discussion-based small seminars led by regular faculty. Hutchins believed strongly in academic freedom and empowered the faculty to engage the students in robust debate.[1] During World War II, the student body declined from about 3,000 to slightly below 1,400, with women in the minority at about 44% of the population. Most of the students came from the Chicago area, with a majority of them choosing to live at home.

For Joanne, everything was new. It was a transformative first year at a great university in a great city. On Monday, September 23, she gathered with her peers in the cavernous and ornate theatre of Mandel Hall, morning sunshine streaming through the Tiffany stained-glass windows. President Hutchins welcomed the new students, reminding them of the challenges they would experience in their upcoming studies. He touched on the dire situation Britain faced in the European war and the uncertainty generated by the newly enacted Selective Training and Service Act. Hutchins finished by expressing his hope that, given the tumultuous times, the students might complete their courses of study and prepare themselves to contribute to a new and better world. Joanne was busy the rest of the week, taking batteries of placement tests, registering for classes, attending social events and a welcoming dinner for all students, settling into her new room in Beecher Hall, and meeting President Hutchins and his wife at a reception for students and parents (which Joanne attended alone).[2]

Joanne was delighted to learn that she was the recipient of a one-year honor entrance scholarship. She enrolled in a set of required core courses: psychology, politics, physics, and astronomy, and remarked that they were "oh, so interesting."[3] The psychology course provided her with basic concepts and the vocabulary to start to analyze her own life and the behaviors of her friends, a practice she continued through life. The politics course, and just living in the city, exposed her to the crooked, machine-dominated Chicago of the Kelly-Nash era, where gambling and organized crime ran rampant. She became entranced with the evolution of the universe from the lectures of Enrico Fermi,

---

[1] Boyer, *University of Chicago: A History.*

[2] *Daily Maroon* 40 (September 11, 1940), 1.

[3] Simpson Papers, 9.2. Narrative by Simpson, January 1994, re: description of undergraduate and graduate college career.

a Nobel Laureate, Walter Bartky, Dean of the Division of Physical Sciences, and Reginald Stephenson, master teacher and author of *Exploring in Physics*. As she reminisced about this in 2000: "If World War II had not come along I probably would have tried to be an astrophysicist. I know many women in the field now. I didn't even consider at the time whether being a woman was important or not, because as an undergraduate at the University of Chicago men and women were treated in the same way. So gender problems didn't even occur to me at that point. I thought I would just go into whatever I was interested in."[4] She said she was "somewhat afraid of boys," but adapted quickly, relishing coeducational classrooms, academic challenges, effective mentoring by her professors, new friendships, and her newfound freedoms.[5] Joanne was one of only five freshmen women elected to the Ida Noyes Council, which planned social, artistic, and athletic events for the entire campus.

Although she had hated the course on physical science in high school, finding it boring to focus on machines and levers, physics became much more relevant now as she continued her adventures in piloting small aircraft, studying meteorology, air navigation, and aircraft structures and engines. It was, as she recalled, a matter of survival: "Those engines became much more exciting after sitting behind the sound of a single one spluttering into silence above the cornfields."[6] Joanne, a civilian pilot, was elected president of the Chicago Flying Club, which flew out of Harlem Airport in Oak Lawn. In pursuit of her commercial pilot's license, she passed both the basic and advanced meteorology exams. She had hit her stride, both academically and socially, in her first semester at college.

A Meeting with Rossby

Joanne was horrified by Hitler's triumphs in Europe and the London blitz. In 1940 her midwestern classmates thought it was none of America's business. But by the late spring of 1942, bad news was arriving daily from Europe and the Pacific about the frightening numbers of US casualties, including several of her male friends and the husbands and brothers of her girlfriends. Joanne was on the verge of volunteering for military duty, a decision that evoked an angry reaction from her mother, but

---

[4] Simpson, "Oral History," by Harper.
[5] Simpson, "Interview," by LeMone.
[6] Simpson Papers, 9.2. Narrative by Simpson, January 1994, re: description of undergraduate and graduate college career; *Daily Maroon* 42 (May 26, 1942), 1.

encouragement from her father. She hatched a plan to leave college and join the WAVES (Women Accepted for Volunteer Emergency Service), where she could get a commission, most likely as a desk clerk or other support role. It was at this critical juncture that she asked her meteorology instructor about taking additional courses. He suggested she visit Carl-Gustaf Rossby in Ryerson Hall (Figure 2.1). Rossby, a gregarious Swede, was the leading atmospheric scientist in the world and chairman of Chicago's Institute of Meteorology.[7] The Institute had already trained two classes of aviation cadets for the Army Air Corps. When the cadets graduated from the nine-month A-course of intensive training in meteorology, they became second lieutenants and were sent around the globe to forecast the weather for military operations.[8]

Figure 2.1 C.-G. Rossby. Credit: University of Chicago Library, Special Collections Research Center.

[7] Fleming, "Carl-Gustaf Rossby."
[8] University of Chicago Library, Department of Special Collections. Department of Meteorology.

Rossby welcomed Joanne into his office to discuss her future. She described her interest in serving the war effort, most likely as a secretary, either in or out of uniform. Rossby, however, thought she could do much better than that. He said, "Forget it, we need science-trained women to serve their country here, after completing the A-course, to teach further A-courses to strengthen the military weather services." He recognized her talent and training in aviation meteorology and told her, "The country needs you."[9] Why throw away your education when you can be trained as a laboratory instructor and finish your BS in meteorology here? Furthermore, the course offered a $65 per month assistantship, which would pay for her education.[10] Unbelievable! Joanne was overwhelmed by the realization that she could serve her country, learn aviation-related skills, finish college, and perhaps finally please her mother while earning her own way and doing something much more interesting than typing or filing.

There was one problem, and it was not a minor one: Joanne did not possess the minimum prerequisites to begin the A-course in September. Rossby immediately ordered his secretary to enroll Joanne in the double-paced summer school to make up for all the deficiencies. She had a near-100 on a pre-calculus comprehensive exam, but no calculus, and no three-course sequence in "sophomore physics." There was no problem getting into differential and integral calculus, but the physics department balked, because the first course, Mechanics, was already finished, and the next two, "Heat, Light and Sound" and then "Electricity and Magnetism," would have to be taken simultaneously. Undaunted, Rossby picked up the telephone and called his friend, physicist Harvey Lemon, telling him that Joanne was a straight-A honor student and that an exception had to be made in the national interest. Lemon said "OK, but she'll certainly fail; no one could bone up on mechanics while taking the double speed physics sequels and double speed calculus at the same time; moreover, the courses have already started."[11]

Her head spinning, Joanne left Rossby's office. She later recalled, "Within ten minutes I was entrained into his orbit."[12] She was now

[9] Simpson Papers, 9.2. Narrative by Simpson, January 1994, re: description of undergraduate and graduate college career.

[10] Turner, "Teaching the Weather Cadet Generation."

[11] Simpson Papers, 9.2. Narrative by Simpson, January 1994, re: description of undergraduate and graduate college career.

[12] Tao et al., "The Research of Dr. Joanne Simpson," 1.

enrolled as a civilian student in the fourth wartime A-course to begin on September 1, 1942. That summer was a nightmare for Joanne, as the classes she had to take challenged her in ways in which she had never been challenged before. She struggled, but the summer was moderated a bit by her aptitude for calculus and her brilliant lab partner, a physics major who soon became her new boyfriend. The B she earned in the electricity and magnetism course represented the first break in her straight A record. Horace Byers, one of the founders of the Institute of Meteorology, noticed her and included an oblique reference to her in his memoirs: "I remember a young lady, Joanne Gerould, now Joanne Simpson, was doing undergraduate work in the Physics Department. We used to see her in the halls and wonder who she was."[13] Byers might have mentioned this to his colleague, Victor Starr.

Teaching Meteorology

As the war-training course began that fall, Joanne, a 19-year old college junior, felt like an outsider. Several of the civilian candidates, all older males, had already earned their degrees in physics or math. On the first day of class, John Bellamy, a course instructor who had a master's degree in nuclear physics, caused quite a stir by leading a group of advanced students in deriving the whole set of the hydrodynamic equations on the blackboard. Joanne was rightfully terrified; she had never heard of the hydrodynamic equations. They seemed to her at the time as difficult as general relativity might appear to a high-school student. The pace of the course was hectic, and the schedule was grueling. Classes began each day at 0800 hours. There were frequent examinations, lots of homework, and required lectures in the evenings. Plotting and analyzing daily weather maps occupied each afternoon. Joanne learned to operate on little sleep, but lived fairly well, sharing a small apartment with two other women. Some of the military aviation cadets, who knew little to nothing about meteorology, were woefully underprepared for the course, and quite a few of them failed. All of the civilians finished, but the seven enrolled women students placed in the top half of the class.

To complicate things even more, Joanne was involved in her first, very serious relationship with an assistant professor at the university.

---

[13] Byers, "Founding of the Institute of Meteorology."

For the first time, if only briefly, she felt lovable. By March 1943, however, he told Joanne that he had found somebody else and did not want
her in his life any more. She was completely devastated and depressed.
Redirecting her energies, she graduated from the A-course ranked
tenth in the class of 210. She received her bachelor's degree in June and,
to her astonishment, notice of her election to Phi Beta Kappa.

After graduation, New York University requested a Chicago graduate to
serve as an instructor in the sixth war course. Joanne volunteered immediately, because, she explained later, she felt so "cracked up," and she just
had to get out of Chicago. At NYU, Joanne stood in awe, mainly fear, of
the faculty who looked askance at her. Yale Mintz and Hans Panofsky were
friendly and supportive, but Athelstan Spilhaus, chairman of the department, made it abundantly clear that he did not want any women on the
faculty. Although they performed their duties well, Joanne and another
female instructor were merely tolerated as part of the war effort.

Joanne returned to Chicago in May 1944 as an instructor with a new
salary of $200 a month. She enjoyed teaching and wanted to make a
career out of it by pursuing her master's degree. She had just turned 21
when she began dating meteorology instructor and graduate student
Victor Paul Starr. He was 14 years her senior and one of the bright rising
stars of the department. Starr impressed Joanne and seemed to offer her
just what she was seeking. He was an accomplished teacher, a seasoned
researcher, and a department insider who took her to dinners at
Rossby's home and nights out with Rossby on the town. It was a whirlwind romance. Dazzled by all the attention, Joanne and Victor married
in June 1944 (Figure 2.2).

They were totally incompatible. She coldly recalled later in life, "If
I had lived with him for a week prior to getting married, I never would
have done it." They started married life on the wrong foot, dealing a
severe blow to Joanne's quest for love and acceptance. The honeymoon
they never took and Joanne's promising career as an instructor ended
on June 13, 1944, the day after her marriage. She recalled, "I was then
abruptly informed that there was a nepotism rule, so only Victor could
be on the faculty, and also that it was time for married women to go
home and mind the mop, since they were only needed to fill 'men's
jobs' during a war that was nearly over."[14] At the time, nepotism rules

[14] Simpson Papers, 9.2. Narrative by Simpson, January 1994, re: description of undergraduate and graduate college career.

Figure 2.2 Joanne and Victor Paul Starr in Chicago, 1945. Simpson Papers, 4.8. Simpson Symposium, February 9-13, 2003.

were major obstacles consistently used by administrators to disadvantage working, married women. Victor knew about this rule, but had not informed her, a deception Joanne found hard to forgive. She recalled the situation later: "At the end of the war, like Rosie the Riveter, we women were supposed to go back to our mops and babies. When about three or four of the roughly thirty war-trained women wanted to go on to graduate school for master's or doctor's degrees in meteorology, our professors were considerably shocked. Some laughed, and a few in key positions were openly hostile."[15]

Work was more than a driver in Joanne's life. Her professional involvement in meteorology was her identity on the one hand, and on the other, served as a spellbinding and largely effective escape from largely dysfunctional family situations and her personal bouts of depression. Joanne hated housework and would do anything to avoid being identified solely as a housewife. The couple rented a shabby and shabbily furnished apartment at 5466 Woodlawn Avenue, about a 15-minute walk to campus. Joanne said that Victor had problems about

[15] J. Simpson, "Meteorologist," 44.

spending money, and to make ends meet she accepted a routine and boring job at the Institute, analyzing seemingly endless numbers of weather charts of the northern hemisphere. In December 1944, three months pregnant, Joanne decided to quit this job and, against all odds, personal and professional, pursue her master's degree in meteorology.

Advanced Study

Although it was co-educational from the start, the University of Chicago catalog emphasized educating "young *men* who can become great scholars or outstanding practitioners in their field." This explicit bias only served to strengthen Joanne's resolve to become a future leader. Rossby had recruited a stellar faculty at Chicago that represented various schools of thought. The department aimed to link theory and practice, expose students to a wide variety of stimulating viewpoints, develop new ways to see meteorology, climatology, hydrology, and oceanography as a unified whole, and support innovative research projects involving cutting-edge instrumentation and new data sets. While these goals were lofty, and the rhetoric was soaring, the realities for a young woman entering the program were discouraging.

The University of Chicago fell far short of implementing Harper's goals of gender equality. To attain a master's degree in meteorology, a student (assumed to be male):

> First, must successfully complete an approved program of study, which is drawn up after consultation with the departmental counselor. Second, *he* must pass a reading examination in German, French, or Russian. Third, *he* must satisfactorily complete a research project which is approved by the department and which is undertaken under the sponsorship of a member of the staff. Fourth, *he* must pass a final examination in the field of meteorology.[16]

But Joanne accomplished a fifth task—one that none of her male cohorts ever could. She became a mother when David Victor Starr entered the world on June 30, 1945. Joanne was increasingly left on her own, with no evidence of a support network around her or record of contact with her family. Her mother Virginia did not visit; she *lent* her money to purchase the baby's crib and other necessities. Joanne never mentioned it, but her brother Daniel had graduated from Boston Latin

---

[16] "University of Chicago, Department of Meteorology."

High School at age 16 and had entered the University of Chicago in 1944. However, there is no evidence of their interaction. She wrote, "I was very unhappy in this period from discouragement by everyone and Victor's lack of interest in anything but his own work."[17] The one exception was sailing on Lake Michigan with Victor, but, since money was tight, she decided to sell her share in the sailboat to pay for graduate tuition.

Joanne was awarded a master of science degree in August 1945, in absentia, for work done on a thesis assigned by Rossby—a mathematical exercise she thought had no physical meaning. Rossby asked her to develop a linear approach to large-scale planetary waves in the upper-level mid-latitude westerly winds. Inspired by the recent discovery of the jet stream, Joanne examined perturbations in a wind field whose velocity increased with altitude. With high hopes, she submitted a paper based on the thesis for publication in the *Journal of Meteorology*, but was severely disappointed when the editor rejected it. Instead, in 1945, under the name Joanne Gerould Starr, she published her first short paper on a different subject: a short note on internal oscillations in ocean currents.[18] She was not proud of her master's thesis and did not retain a copy of it in her files.

Despite Rossby's earlier encouragement and her straight-A record, Joanne was not offered a graduate fellowship. Rossby thought that offering financial support to women was a lost cause and constituted an improper request. He had no use for women in meteorology, and told Joanne frankly that she would look both ridiculous and pathetic if she did not really "make it big" after creating a spectacle of herself and exacting such an unfair sacrifice from her husband and child.[19] Senior colleague Erik Palmén remarked to her as she was preparing her thesis proposal, "It is an ironic tragedy to see the brain of a man in the body of a woman." The graduate student adviser, Michael Ference, encapsulated the chilly atmosphere towards women in the department when he told Joanne bluntly, "No woman has ever earned a PhD in meteorology. No woman ever will. Even if you did, no one would give you a job." Her first reaction was, obviously, to be very discouraged. She thought it over for a few days, then thought, "Well, I'll show him if it's the last thing I ever do." She described herself as a "rather stubborn individual" who responded to

---

[17] Simpson Papers, 1.16. Family history: Simpson's narrative, January 1998, re: marriages, children, and other family from 1940s to 1970s, and captions for family photographs.
[18] J. G. Starr, "Note on Internal Oscillations."
[19] J. Simpson, "Meteorologist," 44.

discouragement with renewed determination to succeed. But it was a hard road, especially without female role models or mentors.[20]

Male professionals at the time erroneously assumed that women went to college to find a husband—and to graduate school to find a more interesting husband. Joanne was strongly determined to succeed at something and avoid finding herself in her mother's position: financially dependent on a husband in an unhappy marriage. She considered leaving meteorology, perhaps for medical school or graduate study in psychology. There were no scholarships or financial aid for women in medical school, and no one would loan her the money, including either of her parents, who were then divorced, nor her husband, who had, for those times, a substantial bank account. She had taken one graduate course in psychology at Chicago, and the department had vaguely promised her some kind of assistantship, but the more she looked at their program and where it might lead with regard to lifelong work, the more her enthusiasm dwindled.[21]

By September, infant in arms, Joanne decided to take on the new challenge of earning a PhD in meteorology. It required two additional years of graduate course work, successful completion of a preliminary examination, a reading knowledge of German and one other foreign language, and a program of research approved by the department. All of her cohorts, save one, were men. Joanne's first PhD supervisor was Rossby, but, echoing his previous sentiment, he did not encourage her because he did not think there was any future at all for women in meteorology. She was trying to emulate the kind of dynamical calculations that Rossby, her husband Victor, and their students were doing, but her heart was not in it, and she did not consider herself very good at it. She did, however, discover something important about her supervisor that inspired her and served her well in the future. In addition to being a theorist, Rossby strongly believed in being a close observer of nature. He recommended that all PhD students in meteorology become either private airplane pilots, glider pilots, or sailors.[22] Joanne did just that.

[20] Simpson Papers, 9.2. Narrative by Simpson, January 1994, re: description of undergraduate and graduate college career; Simpson Papers, 9.3. Work Notebook #1, development of PhD dissertation, August 1947–June 1949; Raymond and Carlson, "My Daughter the Scientist," 24.

[21] Simpson Papers, 9.3. Work Notebook #1, development of PhD dissertation, August 1947–June 1949.

[22] Simpson, "Interview," by LeMone.

Joanne remembered this period as "one of the darkest of my life." Her marriage was deteriorating. Victor was not an attentive partner, a loving father, or a good provider. The couple separated. Struggling to make ends meet, Joanne looked, without success, at all universities and junior colleges in the Chicago area for a teaching or research job. Finally, in 1946, James Thompson, head of the physics department at Illinois Institute of Technology, telephoned to ask if she could teach beginning physics as a part-time instructor. Joanne agreed immediately. She suggested offering a course in elementary meteorology, which, she argued, all engineering students should have, since the weather was so important to the structures and systems of the built environment. Thompson agreed, on the condition that enrollments exceed twenty students.[23] The new job helped lift her out of depression, although she rarely had a free moment, and the starting salary was less than she needed to make ends meet. She was living alone, raising her son, pursuing her PhD, and teaching a full range of undergraduate physics courses, including mechanics, electricity and magnetism, special relativity, and meteorology.

Joanne did not aspire to turn her junior and senior engineering students into weather forecasters or research meteorologists. Instead, she emphasized understanding the atmosphere and its relevance to engineering.[24] A dam builder needs information about the rainfall characteristics of a watershed; an airplane designer must know about aerodynamic stresses, icing conditions, and severe weather; the designer of an air terminal too must consider a plethora of weather factors, including prevailing winds and fogs. In the laboratory, students measured humidity by swinging a psychrometer, counted the revolutions of an anemometer, and identified fronts and air masses on weather charts. They launched and tracked radiosondes and peered into the fog in a home freezer as they reenacted Vincent Schaefer's cloud-seeding experiments with dry ice. They especially enjoyed the field trips and hands-on experiences offered on Saturdays as extra course work. On an outing to the University of Chicago they watched the swirling colored fluids in the rotating dishpan apparatus, dimly recalling that they had once seen the equations of geophysical hydrodynamics. In the final week of the course they visited the US Weather Bureau forecast office to observe the meteorologists,

[23] Simpson Papers, 8.3. Simpson's narrative re: teaching meteorology class at Illinois Institute of Technology.
[24] Ibid.

surrounded by weather maps and charts, working under severe time constraints amid the clacking teletypes and clanging telephones. Her students were on the way to becoming weather-savvy engineers, ready to interface with the weather enterprise.[25]

At Chicago, Joanne was surrounded by an all-male meteorology faculty, including her estranged husband Victor. There were no childcare facilities, and she was not even allowed to bring infant David into the department library. Rossby told her it was "unseemly and undignified—what if the Dean should walk by?" In spite of all the obstacles, Joanne persevered in pursuit of a PhD, which still appeared to be her best chance to remain in university teaching, or perhaps move into another field if necessary. Rossby moved back to Sweden in 1947, which relieved some of the pressure on her. The new department chair, Horace Byers, seemed somewhat less vocally opposed to a woman trying to get a PhD. That year, Victor and Joanne divorced, and Victor voluntarily relinquished custody of their son David to his mother. Illinois Tech promoted Joanne to full-time instructor, allowing her to reach one of her goals: being able to pay her living expenses, but with time and money in short supply, progress on her degree suffered. Still, she was able to complete her preliminary course work in the spring of 1947.

Although she was still a PhD candidate, the American Meteorological Society asked Joanne to author the section on "Women in Meteorology" for its 1947 employment survey, *Weather Horizons*.[26] Here she looked comprehensively at how women could continue to achieve in meteorology, and asked whether the small, albeit noticeable, boom in job positions had been merely a wartime phenomenon. The US Department of Labor's report on the outlook for women in science cited her conclusion that opportunities will continue to be limited in number. She predicted, in a hopeful exercise in self-prophecy, that a few enthusiastic young women with excellent backgrounds in physics and mathematics, preferably with training not only in meteorology but also in a related scientific or industrial field, will find opportunities. For such women, "the horizons in weather are wide, for those with vision enough to see beyond the obstructions. Ceilings and visibilities are almost unlimited for those women with the enthusiasm and initiative to make their

---

[25] Simpson Papers, 8.3. "The Engineer and the Weatherman," *Illinois Tech Engineer* (Oct. 1948).
[26] J. G. Starr, "Women in Meteorology."

own opportunities, and the character to stick out the hardships."[27] Joanne's writing about a future she could only anticipate revealed her confidence and her adherence to a kind of "first wave" feminism that assumed social change could be achieved if (educated white) women were given the chance to prove they were just as capable as men.[28] Joanne was threatening tradition by not leaving the profession after the war emergency and by aiming for positions that were looked on as male prerogatives. Speaking from personal experience, Joanne warned that prejudice against a woman's holding a professional position appeared to be greater in meteorology than in the other sciences.[29]

The Tropical Turn

The event that changed Joanne's research trajectory was professor Herbert Riehl's course in tropical meteorology, the first one ever offered anywhere, and the basis for his important book, *Tropical Meteorology* (1954). Riehl (Figure 2.3) spent the first two weeks discussing convection in the tropics. He focused on recent observations by Jeffries Wyman and Al Woodcock of cumulus clouds and their surrounding environment in the trade winds north of Puerto Rico. The expedition, sponsored by the Woods Hole Oceanographic Institution, collected profiles of temperature, humidity, and turbulence from the sea surface to 3,000 feet using an array of ships and an instrumented Catalina "flying boat," a venerable and reliable military airplane first developed in the 1930s that saw extensive service in World War II.[30] Woodcock, a consummate naturalist, was able to visualize updrafts and turbulence by observing the clouds, the response of the airplane, and the soaring behavior of herring gulls. In weak winds and in certain locations, the gulls were able to spiral upward under clouds without flapping their wings. In strong winds, they soared in straight lines along the rising portion of roll vortices, which typically define cloud streets in the trade wind regime. Upon examining these results, the noted oceanographer Henry Stommel remarked, "Well, these clouds aren't obeying the classical equations. They are getting

[27] US Department of Labor. "Outlook for Women," pp. 7–39; J. G. Starr, "Women in Meteorology," 29.

[28] Hewitt, *No Permanent Waves.*

[29] Noble, "Joanne Simpson, Meteorologist."

[30] Simpson Papers, 9.1. Writings: "The Riehl–Malkus Collaboration on Tropical Meteorology and Hurricanes," Memoir by Joanne Simpson, 1998.

Figure 2.3 Herbert Riehl. Source: Department of Atmospheric Science, Colorado State University.

diluted by air from the outside."[31] Stommel had identified the phenomenon of cumulus entrainment in which clouds exchange air with their drier surroundings.[32] This opened up a completely new area of research in tropical meteorology. When, in class, Riehl lucidly described trade-wind clouds, the data this expedition obtained, and Stommel's new theory, Joanne suddenly saw the proverbial light-bulb come on above her head, and shouted "EUREKA! This is what I want to study for the rest of my life."[33]

A fortuitous occurrence gave Joanne her start on a dissertation project. While visiting Cambridge, Massachusetts, in July 1947 she was reacquainted with meteorologist Bernhard Haurwitz, then at MIT, who was dating her mother. He asked Joanne, "What are you doing in

[31] Simpson, "Oral History," by Harper.
[32] Stommel, "Entrainment of Air into a Cumulus Cloud."
[33] Simpson Papers, 9.1. Writings: "The Riehl–Malkus Collaboration on Tropical Meteorology and Hurricanes," Memoir by Joanne Simpson, 1998.

meteorology?" She said she was really excited by what the Wyman–Woodcock data were revealing about clouds in the tropics. He replied, "Well, that's very interesting. I'm in charge of a project down at Woods Hole to do further work on the data from that expedition. Would you like to come down for a few days and visit and see what's going on?" Virginia volunteered to take care of two-year old David for several days, while Joanne visited Woods Hole. Joanne found the data interesting and the work environment welcoming, so she asked Haurwitz if she could visit again the following summer. He said, "Sure."[34] During this time, Joanne also met photographer Claude Ronne, who became central to the next chapter of her life.

Joanne returned to Chicago and asked Herbert Riehl if he would be her major professor and supervise her thesis on tropical meteorology. As Joanne recalled years later, "He was the kind of guy that would make a pass at anything in skirts, so I thought I would take advantage of that and ask him if I could be one of his graduate students." Riehl assented, adding with false modesty, "Well, I don't know anything about cumulus clouds, but if you don't mind that, sure."[35] Actually, Riehl learned about cumulus clouds very fast, because he soon suspected that they might play a crucial role in the behavior of the Pacific trade wind layer. The duo enjoyed a long and productive research collaboration.

Herbert Riehl (Joanne called him Herbie) treated his students and associates as part of his family, so it was no surprise that he took Joanne under his wing, both as a mentor and as a friend. Joanne had taken several graduate courses at Chicago in psychology, culminating in a course on abnormal psychology. In this course, a great deal of discussion concerned hypnotism, how it is induced, and what role it can play in the treatment of psychological disorders. One of the main points was that an individual could recall suppressed and apparently forgotten parts of his/her life under hypnosis. The professor wanted to give a demonstration of hypnosis for the class to show how material in the subconscious could be brought out. He asked for volunteers, interviewed them, and chose Joanne, who seemed to be fairly normal and only moderately suggestible. When Joanne told Herb about this in a fairly off-hand way, he got quite concerned, almost upset. He insisted on finding out the place and time of the next demonstration so that he could be there in

[34] Simpson, "Oral History," by Harper.
[35] Ibid.

the audience. He said his European background led him to believe that hypnotism was pretty risky and the subject might reveal all sorts of skeletons in the closet, real or imagined. Actually, if Herb had not attended the demonstration, Joanne never would have known what happened, since the professor apparently said she was to forget it all afterward. What the professor did in the demonstration was to take her back to third grade. He asked for the name of the teacher and information about her fellow pupils (which, of course, she did not remember when conscious). He also asked her to write her name and her teacher's name on the blackboard. When she came out of the hypnosis, Joanne saw both names written in the childish printing that she actually used in third grade but had long since abandoned. Greatly relieved that no skeletons had surfaced, Herb took Joanne to lunch and talked about the funny episode for some weeks afterward. She was impressed that he cared enough about her to take time to be concerned and to attend the demonstration.[36]

Herb and Joanne had lots of discussions of books they were reading not related to meteorology. The works of Dr. S. I. Hayakawa on semantics were stimulating—in particular, *Language in Thought and Action*, in which he wrote, "In a very real sense, people who have read good literature have lived more than people who cannot or will not read. It is not true that we have only one life to lead; if we can read, we can live as many more lives and as many kinds of lives as we wish." Dr Hayakawa was then on the University of Chicago faculty and gave night courses for adults in downtown Chicago. Herb and Joanne decided to sign up for the course and travelled together on the IC electric train (neither of them had access to a car). This turned out to be a fascinating experience, because Hayakawa used interaction with the students and their experiences as the primary material in the course. The course was over at 8 pm, and Herb and Joanne got into the habit of having dinner together afterward at his favorite Viennese restaurant in the Loop area. Anyone who knew Herb knew about his connoisseur-class expertise in martini cocktails. At these dinners he got Joanne affectionately acquainted with the Gibson martini (very dry, straight up, with a cocktail onion instead of an olive).

They occasionally talked about meteorology at these dinners, but mainly Joanne learned about Herb's life: his Jewish mother and flight from Germany in 1933, his sojourn in England, his adolescence in the US, his work on Wall Street, and briefly, as a script writer in Hollywood,

[36] Reiter, "Herb: Personal Recollections," 6.

his war training as a weather cadet at NYU, and his work at the Institute of Tropical Meteorology in Puerto Rico.[37] Herb was only eight years older than Joanne, and she recalled taking advantage of his flirtatiousness: "He tried to make a pass whenever he could, but I managed to resist just enough to keep him interested. And oh, we got to be really good friends and colleagues. I think both of us did our best work in working with each other, so it was worth using some slaps in the face every so often. If he'd get too bad, I'd go out and slam the door, which apparently all the other graduate students thought was hilarious. We didn't even know they were paying any attention."[38]

Joanne's first research notebook reveals her struggle to find a dissertation topic. Her former husband Victor, who she considered to be one of her best teachers and mentors, had impressed on her the importance of keeping a notebook to record, almost daily, her thoughts on developing research. Starting in 1947, she made regular entries for a decade until the rise of computer modeling reduced the utility of this practice. These journals document her turn away from theoretical meteorology, which she was not good at and did not find very interesting, and her new fascination, supported by Herb, with field programs to study meteorology in the tropics.[39] Still, not everything went smoothly. The unsuccessful part of her thesis, influenced by Rossby and Starr, was highly theoretical. It involved attempts to treat waves in the tropical easterly winds as perturbations on a basic current, so that linear mathematics could be used. The *Journal of Meteorology* rejected her manuscript on this work, not only utterly, but with extremely disparaging remarks by the reviewer, leading Joanne to wonder if there was prejudice involved against an unknown woman, or just a low opinion that would have been expressed had the writer been a much better known man.[40] Although this part of her thesis did not produce any significant results, it was a valuable learning experience, serving to redirect her interests towards observational projects, the analysis of new data sets, and unexamined topics.[41]

In April 1948, Joanne married Willem Van Rensselaer Malkus, a part-time instructor in the physics department at Illinois Tech and a graduate

[37] Ibid.
[38] Simpson, "Oral History," by Harper.
[39] Simpson Papers, 9.3. Work Notebook #1, development of PhD dissertation, August 1947–June 1949; LeMone, "What We Have Learned about Field Programs."
[40] Simpson Papers, 9.1. Writings: "The Riehl–Malkus Collaboration on Tropical Meteorology and Hurricanes," Memoir by Joanne Simpson, 1998.
[41] Ibid.

student at Chicago studying under Enrico Fermi. She changed her name to Joanne Starr Malkus, probably keeping the middle name Starr for professional reasons. Willem legally adopted Joanne's son and changed his name to David Starr Malkus. A second son, Steven Willem, was born June 13, 1950, the same week that Willem received his PhD in physics from Chicago. Joanne recalled her second pregnancy and Riehl's cigar-smoking vice:

> Those who knew Herb Riehl in the Chicago days know that he loved nothing more than a good cigar. He smoked cigars regularly, both in the work place and everywhere else he went. On the other hand, I associated cigar smoke with seasickness experienced as a child. When I was pregnant with my second son in 1950, my nausea-prone condition worsened. That was the period when Herb and I were doing our work on the trade-wind circulation, the heat balance, and the maintenance of the trades. Herb stopped smoking cigars. I was amazed and asked him why he did that. He said that of course the reason was to keep me comfortable. I was very touched because I had not asked him to make such a sacrifice. There was no smoking allowed in his laboratory until after Steven's birth.[42]

In the summer of 1948, Joanne returned to Woods Hole, where she worked on the Wyman–Woodcock cloud data, watched time-lapse films of cumulus clouds near Woods Hole, and developed a mathematical theory of how drier air mixes with a cloud (entrainment) in response to wind shear, demonstrating the asymmetrical interactions of clouds with their environment. Results indicated that clouds that moved faster than their surrounding environment left behind some of their moisture as cloud droplets, or as Joanne summarized it: "The clouds shed moist air and thereby moisten the environment."[43] It was new and fascinating work for her, given her lifelong interests in aviation and clouds. This work comprised the successful part of her PhD thesis and set the stage for her future research on the tropical atmosphere. By the fall of 1948 she had worked out her thesis plan in essentially final form and had no problems getting her degree after drafting it. The oral defense went surprisingly well: "Professor Riehl had spent the preceding morning discussing various aspects of the work with her and then led off by asking her precisely the same questions they had discussed previously. Professor Wouter Bleecker, one of the leading European cumulus cloud scientists, participated fairly vocally, but all went smoothly.

[42] Reiter, "Herb: Personal Recollections," 8–9.
[43] Lewis et al., "Herbert Riehl."

Professor Byers rather liked the thesis."[44] Her work on tropical convection constituted a total departure from the work of any of her mentors and the beginnings of her own original thinking about cumulus clouds and their relationships to the air in which they are embedded.

On a warm sunny day in early June 1949, Joanne received her PhD formally from Chancellor Robert Maynard Hutchins, who gave a stirring graduation address denouncing McCarthyism as an attack on the First Amendment. As she marched up the aisle, the piping voice of little David, nearly four, rang out through Rockefeller Chapel, "That's my mommy!" and everybody in Rockefeller Chapel broke up laughing. Joanne likely paused for a moment on her way into and out of the chapel, since grandfather "Gubby's" wrought-iron manufacturing company had made the door handles and fixtures. In 1949 Joanne was promoted to tenure-track assistant professor at Illinois Tech at the then generous salary of $3,700 for a nine-month contract.

After she received her PhD, Joanne published her first few papers on cumulus clouds, including one on the effects of wind shear and one with colleagues from Woods Hole on the vertical structure of the atmosphere over the Caribbean Sea.[45] Later that year, Rossby returned to Chicago for a visit. He greeted Joanne in a jovial, avuncular manner, saying that her work on cumulus clouds comprised a fine field for a "little girl" to work on, because it was not important enough to interest real meteorologists. The recipient of a PhD is normally accorded a warm welcome into the scholarly world, but it took eight years until Rossby acknowledged Joanne's membership in the class of "real meteorologists." Even Victor was able to admit, the last time he saw Joanne in about 1950, that she had become "a damned fine meteorologist."[46] Ever seeking validation, she considered this "one of the most treasured compliments I ever received," since Victor was personally very bitter and said, almost in the same breath, that he never wished to set eyes on her again.

---

[44] Simpson Papers, 9.1. Writings: "The Riehl–Malkus Collaboration on Tropical Meteorology and Hurricanes," Memoir by Joanne Simpson, 1998; Malkus, "Certain Features of Undisturbed and Disturbed Weather."

[45] Malkus, "Effects of Wind Shear"; Malkus et al., "Vertical Distribution of Temperature and Humidity."

[46] Simpson Papers, 9.2. Narrative by Simpson, January 1994; re: description of undergraduate and graduate college career.

3

Woods Hole

Nothing in life that is worth achieving is gained without compromise, without self-discipline, or without effort and some or many heartaches along the way.

JOANNE MALKUS[1]

Joanne's career took off at the Woods Hole Oceanographic Institution where she worked as a research associate for four summers, from 1948 to 1951, and then as a full-time employee until 1960.[2] At the same time, her turbulent emotional life, driven by a life-long quest to find love and be lovable, became, if anything, even more turbulent. Her relationship with second husband Willem deteriorated as her all-encompassing and not-so-secret love affair with Woods Hole's chief research photographer Claude Ronne blossomed. Joanne and Claude first met in the summer of 1947 and engaged in a brief but memorable conversation before she was hustled away by her host, Bernard Haurwitz. Joanne would later worked with Ronne, who had photographed the 1946 Wyman-Woodcock expedition to the tropics.[3] But her most vivid recollection of the encounter was their discussion of a Shakespearean play, Richard III, then being featured at a nearby summer theatre, and how the portrayal of Richard deviated from the facts of history. Joanne later described the meeting as "truly love-at-first sight,"[4] at least from her perspective. Although her first impression was strong, nothing much came of it, since she thought Claude was unattainable, a man too handsome, cultured, and witty to pay much attention to her. In subsequent summers, she enjoyed her visits, especially catching up with Claude, who helped her interpret the cloud films at the center of her studies of cumulus entrainment. She also admired the leadership style of Woods Hole

---

[1] Simpson Papers, 1.7. Personal Diary III: 304, Sept. 4, 1954.
[2] "Joanne Malkus Simpson Papers, 1951–1964." Woods Hole Oceanographic Institution.
[3] Woodcock and Wyman, "Convective Motion in Air over the Sea."
[4] Simpson Papers, 1.6. Personal Diary I: Introduction by Joanne Simpson, 1999.

First Woman: Joanne Simpson and the Tropical Atmosphere. James Rodger Fleming,
Oxford University Press (2020). © James Rodger Fleming.
DOI: 10.1093/oso/9780198862734.001.0001

director, Columbus Iselin. His attitude was "Anybody can do what they want to as long as they spend some of their time doing things for the company. That is the way he'd say it."[5] Of course, Woods Hole was a wonderful place to be, especially in the summer.

In 1951, based on her positive research experiences there and seeking a fresh start, Joanne, a twice-married mother of two with PhD in hand and second husband in tow, left her job at Illinois Tech. She set out into uncharted waters for women in science, determined to continue her research and make her mark. She moved with the family to Woods Hole to take up an appointment as a research meteorologist, to provide a better environment for raising the children, and in large part, hoping to deepen her relationship with Claude and follow her heart. Joanne referred to her husband Willem as a "brilliant, but very arrogant man who offended a lot of people."[6] Woods Hole hired him as a research associate to examine theoretical problems related to turbulence and fluid mechanics. Columbus Iselin had stepped down as director of Woods Hole by then. His replacement, Edward ("Iceberg") Smith, a forty-year veteran of the Coast Guard, remained under Iselin's influence and maintained a laissez-faire environment for research.[7] Joanne and Willem each earned about $4,500 a year, which seemed like a lot at the time.[8] The couple lived in a small house in Falmouth and shared a common interest in sailing the waters off Cape Cod.[9]

Flying

In 1951 the Woods Hole meteorology team received a PBY-6A airplane through the Office of Naval Research (ONR) as a loan from the navy.[10] At the time, Woods Hole had a rule against women going on the oceanographic ships—a rule they tried to apply to the airplane. Max Eaton and Gordon Lill, both on the ONR staff, objected strenuously, saying "No Joanne, no airplane," so they had to allow her to fly. Being the first

[5] Simpson, "Oral History," by Harper.

[6] Ibid.

[7] Simpson Papers, 2.10. Clippings: "Beginning of research career in atmospheric sciences—Joanne Malkus Simpson, 1953–1964, early recognition, Imperial College, London-Guggenheim Fellowship-clippings."

[8] Simpson, "Oral History," by Harper.

[9] "Willem Malkus, professor emeritus of mathematics, dies at 92."

[10] PBY (Patrol Bomber Y built by Consolidated Aircraft) a US Navy medium to heavy twin amphibious aircraft used for maritime patrol, water bomber, search and rescue, and research meteorology.

and only woman on a flight crew was a "lonely job." Some military facilities would not let her onto the base, and of course, the plane was small and the flights were long, so Joanne often had to go fourteen hours between rest stops.[11] It took about a year to outfit the airplane, place strain gages all over it, and mount an accelerometer at its center of gravity. Seeking to collect data comparable to earlier campaigns, they installed instruments to measure temperature, humidity, and pressure similar to those used by the Wyman-Woodcock expedition. They also added a crude sensor to try to measure liquid water. The PBY was a very slow-flying airplane with a stall speed of about 60 miles per hour and a ceiling of about 7,500 feet. Similar to Woodcock's observations of sea birds, the airplane could be used as a sensor of vertical air motions as it climbed and fell in flight. It was ready for the first field program, in Puerto Rico in June 1952 (Figure 3.1).[12]

Figure 3.1 Joanne's first research aircraft, the Woods Hole PBY, provided by the Office of Naval Research. Left to right: Ken McCasland, instruments, Claude Ronne, photographer, Joanne Malkus, chief scientist, Clayton Kiernen, radio operator, Dick Fournier, co-pilot, Larrs Rose, engineer, Norman Gingras, pilot. Simpson Papers, 455, Photos. PD-9.

[11] Simpson, "Interview," by LeMone; Simpson, "Oral History," by Harper.
[12] Malkus, "Some Results of a Trade-Cumulus."

Working out of San Juan, Puerto Rico, the team took measurements in and around cumulus clouds that were in widely differing phases of their life cycles. They carefully documented one cloud at a time using their instruments and with photographs taken by Claude. Joanne used the data to calibrate a model of the formation and successive growth of larger clouds from several smaller ones. It was part of a first attempt to make a one-dimensional cumulus model to see if the clouds were lining up along tropical "fronts" that were driving them to greater heights, or whether they were driven by positive buoyancy of their own. It was very revolutionary at the time to think that the ascent of undiluted bubbles provided the building blocks of cumulus clouds.[13] Joanne's studies of heated and, in the case of Puerto Rico, rugged tropical islands led to a number of articles and conference presentations examining how small perturbations on the trade winds support and might possibly enhance the development of clouds. Foreshadowing her later involvement in experimental meteorology, she wrote, "The very delicate equilibrium governing the trade-wind air stream raises the possibility of artificial attempts to alter it in limited areas."[14]

In her first project, Joanne examined the effects of islands on the flow of air near the ground. She worked with Andrew Bunker at Woods Hole to study how the heating of a flat island, specifically Nantucket, could lead to the formation of cumulus clouds. Max Eaton, head of the Office of Naval Research, a major source of funds at Woods Hole, liked the project and provided a grant of $5,000, which Joanne used to take measurements and work on a mathematical model of the phenomena together with Melvin Stern, her first graduate student at Illinois Tech.[15] Joanne also photographed clouds over St Croix, Puerto Rico, the Bahamas, and elsewhere. In the spring of 1953, Imperial College conducted a tropical boundarylayer experiment in Anegada—a small, flat, coral island in the British West Indies. The team from Imperial College sent up pilot balloons every five minutes to investigate low-level wind profiles and turbulent mixing. Joanne, accompanied by Herb Riehl, brought the Woods Hole PBY aircraft to fly over and near the island, and stationed Claude on the ground to make quantitative time-lapse

---

[13] Simpson, "Interview," by LeMone.

[14] Malkus, "Effects of a Large Island"; Malkus, "Tropical Rain Induced."

[15] Malkus and Bunker, "Observational Studies of the Air Flow over Nantucket Island"; Stern worked for a time at Woods Hole, earned his PhD at MIT, and went on to a distinguished career in oceanography.

Figure 3.2 Anegada 1953. Photographing clouds in Anegada in 1953 during an Imperial College–Woods Hole cloud survey. This cartoon depicts Joanne making plans to photograph clouds from the island and from the PBY. In an otherwise black-and-white line drawing, the red pilot balloon matches Joanne's red shorts. Simpson Papers, 3.7. Lecture slides (presentation was canceled) re: life of Simpson as scientist, 2004.

pictures of the clouds. Claude was able to photograph distant cumulonimbus from Anegada while the PBY was flying overhead (Figure 3.2).

During the flights over Anegada, Joanne recalled, "Riehl held on to my shorts as I leaned far out of the open blister to get a fine series of pictures with the old reliable Speed Graphic camera. This fortuitous coincidence enabled us to do quite accurate photogrammetry on the clouds."[16] On the occasion of her thirtieth birthday, the flight crew presented her with a cartoon (Figure 3.3) commemorating their adventures in the air.

I was always hanging out of the plastic blister of the PBY to take pictures. With the wind rushing by, someone always had to hold me by the shorts. All the liquor bottles going overboard represent what happened to a

[16] Simpson Papers, 9.3. Work Notebook #1, development of PhD dissertation, August 1947–June 1949.

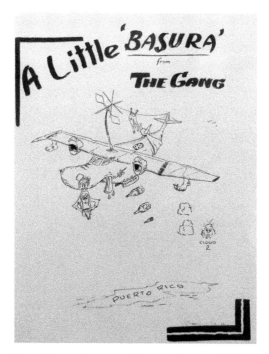

Figure 3.3 A Little "Basura" cartoon, Simpson Papers, 455, Photos. PD-9.

Woods Hole Oceanographic Research vessel that was caught smuggling liquor back to the States—there was a terrible fuss, cars confiscated, lawyers, etc. The aircraft scientists NEVER brought back more than the allowed quota. I was Chief Scientist on the aircraft and, despite the difficulties I had because I was a woman, that was one area where the guys paid attention to me and were later very glad that they had.

The word "Basura"—garbage in Spanish — has a special reason for being in the picture. One day, we were studying the cloud formations over the island of Puerto Rico. I was keeping notes of our location to locate the pictures I was taking and also to reconstruct the whole mission afterward. I asked over the Intercom "What is that village down there below us?" and Ken McCasland, who had all the maps, answered, "That is the little town of Basura". I faithfully wrote down in my notebook "2:10 p.m. town of Basura below us." Later back at the Naval Base in San Juan and walking the streets, I saw trashcans labeled "Basura." I said, "How could someone have brought all those cans from Basura to here?" and everyone else broke up in loud laughter. I suddenly realized that Basura was the Spanish word for garbage and that that sly fox Ken McCasland had pulled one of his best tricks. That evening, I bought his first drink at the officers' club. This was no great reward, since Martini's and all other drinks sold there for 25 cents. Those were very happy and nostalgically remembered days.[17]

Imperial College

In 1951, Herb Riehl and Joanne Malkus received an invitation to visit Imperial College in London to give a presentation to the Royal Meteorological Society on their joint study of the trade winds. Riehl had declined the invitation, so Joanne, who was deeply impressed by the meteorologists at Imperial College, decided to go alone. Her Woods Hole connection to the navy allowed her to travel for free in the cargo hold of a military transport. This presentation—the first fruits of the Riehl–Malkus collaboration—went well. She presented vertical cross-sections of temperature, moisture, and wind motion measurements of the tropical atmosphere northeast of the Hawaiian Islands. The data showed that the trade-wind inversion was not a discontinuity separating an upper dry and lower moist layer, but a permeable boundary through which downward transfer of mass takes place. An imbalance of heat and moisture through the layer revealed a net heat export in the form of latent heat, which served to erode the trade inversion as the winds flowed downstream to Hawaii.

Riehl and Malkus's analysis showed a gradual increase in the height of the trade-wind inversion despite continuous divergence and sinking of air columns. This indicated that a continuous influx of mass was occurring across the inversion. To use Joanne's term, the tropical inversion was a "leaky wall," and associated with it was the formation of cumulus clouds that pushed into the inversion and lifted it. Joanne pointed out what others had failed to see: that cumulus clouds forming around the trade wind inversion were not just epiphenomena, but were transporting something. She worked out the mathematical and physical basis for how these clouds were raising the height of the trade wind inversion, wetting the cloud layer, and making it deeper. In essence, clouds are a key component of the atmosphere's engine and give heat from the ocean surface a way to escape to higher levels and ultimately

---

[17] Simpson Papers, 3.9. Narrative (by Simpson): "at Woods Hole Oceanographic Institution. Summer 1947–1950" and "Pacific Cloud Hunt."

to be transferred northward.[18] The clouds revealed the dynamic process. As the trade wind cumuli pushed into the stable layer at 2–3 km elevation, their tops were sheared off. These so-called "cut-off' towers" transfer mass downward, redistributing moisture and momentum in the vertical. Henry Stommel first used the word "entrainment" to describe convective processes in which clouds exchange air with their drier surroundings, but Joanne certainly made it memorable, since she based her cloud model on the concept and demonstrated its widespread effects.

Joanne found the experience at her talk in London quite intimidating, especially the discussion period. At the time, she did not realize that it was British tradition to attack the guest speaker, or at least offer further insights and challenges. Two senior British colleagues, Richard Scorer, who had invited her, and F. H. Ludlam, a specialist in cloud physics, jumped right up and said they disagreed with everything she said and that entrainment was "rubbish." They attributed the dilution found on cloud penetrations to either instrument error or to the effect of the aircraft on the clouds. They proposed, instead, a ' "bubble theory of convection" ' in which the bubbles remain undiluted at their cores and shed their outsides into a following turbulent wake. In reality, however, they liked her and she liked them. Their working relationship was solidified afterward over several pints at a local pub. The results of this research appeared in the *Quarterly Journal of the Royal Meteorological Society*.[19] Joanne also wrote popular versions of her studies on trade-wind clouds for the British journal *Weather* and for *Scientific American*.[20] This work was important because, for the first time, Joanne was able to show that cloud processes play a significant role in a major component of the large-scale circulation of the atmosphere.[21]

Joanne and Claude

In 1952, Joanne Malkus and Claude Ronne began an intellectual and romantic affair that lasted for decades. They flew on missions together in the Caribbean and the Pacific, passed notes across the airplane's aisle,

[18] Simpson Papers, 9.3. Work Notebook #1, development of PhD dissertation, August 1947–June 1949; Lewis et al., "Herbert Riehl."

[19] Riehl et al., "North-East Trade of the Pacific Ocean."

[20] Malkus, "Aeroplane Studies of Trade-Wind Meteorology"; Malkus, "Trade-Wind Clouds."

[21] Malkus, "Recent Advances in the Study of Convective Clouds"; Malkus, "Slopes of Cumulus Clouds."

danced, listened to classical music, talked all night, and maintained a unique, lifelong relationship unlike any she had experienced. Joanne documented their affair in three hand-written diaries, notes, and letters, made available in 2015, in which she recorded, with remarkable candor, her thoughts and feelings about Claude.[22] Her professional goals at the time involved establishing her scientific reputation as a research meteorologist at the cutting edge of observation, flying in the tropical atmosphere to gather data to help her understand and model the behavior of clouds.[23] Her personal goals included understanding herself and Claude and learning how to be lovable. Joanne considered taking a "quasi-scientific approach" in the journal based on her psychology courses, to analyzing their relationship, but rejected that methodology: "A Freudian psychiatrist would quickly put his finger on what I also would agree with him is the focus and crux of the matter, but in so doing the perspective is distorted and the spirit dead. A human being is more than his skeleton, his conditioned reflexes, and his complexes, just as a concert is more than notes, harmonics, and equations contained in tomes on the physical laws of acoustics."[24] Claude was fun to be with, but exemplified qualities of vulnerable narcissism, for example, in his demonstrated lack of empathy and sympathy for others. He was also introverted and anxious, afraid to take risks, in fear of failure. He was Joanne's colleague, but he was not an accomplished scientist. Joanne's purpose in writing the journals was not to change Claude's behavior, but to critique it and allow him insight for personal growth if he so chose to use it in that way. They were soulmates, but fragile and vulnerable ones at that, on a rollercoaster of feelings that lasted several decades: "What hurts you hurts me, and what makes you happy makes me happy—thus I quite naturally want to do whatever I believe will make you happy, and this is quite natural and there is no need to search for an ulterior motive…I felt so accurately every one of your emotions that the conflict was tearing me apart also."[25]

Their mutual attraction was obvious. Joanne wrote in her diary: "Some of the best moments of my life have occurred in your company

[22] Simpson Papers, 1.6–1.9. Diaries (3), letters, and notes between Simpson and her lover/friend/coworker, 1950s–1984.

[23] Simpson, "Interview," by LeMone.

[24] Simpson Papers, 1.6, Personal Diary I: 115, May 27, 1953; Simpson Papers, 1.16–1.17. Family history: Simpson's narrative, January 1998, re: marriages, children, and other family from 1940s to 1970s, and captions (and extended captions) for family photographs.

[25] Simpson Papers, 1.6. Personal Diary II: 249–51, April 5, 1954.

and were either created or greatly enhanced by your participation. These incidents range from ridiculous to glorious."

> It was a deliciously funny experience to see you, the paragon of beautiful manners, sitting in shorts on a gunner's seat in a littered blister, guzzling sardines in the manner of the Harvard undergraduate and the goldfish. I still remember how hungry, hot, and dirty we were as the oily sardine juice ran down our fingers; how our heads ached from the noise at the same time as we were howling with the giggles . . .[26]

Joanne recalled the evening the PBY was returning to San Juan after a long day of research flying. Multicolored thunderclouds lingered in the sunset. As they approached Roosevelt Roads airfield, banking low around the Morro Castle with the harbor and buoys below them, the lights of San Juan were just beginning to flicker on. Taking her headset off and running her fingers through her short-cropped hair, her gaze turned to Claude across the aisle. She was feeling the stirrings of love and the desire to be close to him.[27]

That a woman had joined the flight crew, as a Principal Investigator no less, was a breakthrough in scientific research, but her presence had also changed the personal dynamics of the team. This is quite clear in her diary entries later in November 1952, when the PBY flew to Bermuda for research, and her romance with Claude was heating up:

> It is indeed strange, this time, to be sitting on my gunners' seat in the jouncing, cluttered blister glancing across at your face, turned in profile to gaze over the sea toward the receding coastline of Bermuda. I look at your face and write to you. Again, the days of intense contrast have passed. You have showed me the blue-green inlet among the rocks that I described to you here weeks ago as epitomizing what we have been. At moments on this trip we have been as close to one another as two complex and troubled humans can be. At times, when we paced up and down the midnight streets of Hamilton under the lighted clock, and twice en route, I have breathed your every breath and my heart followed each one of your heartbeats. The thoughts and fragments passing through your head were communicated with the clarity of the water in our cove, through which every ripple on the sand below was delineated. Then, there were times when the communion ceased, completely as the dead interphone on the old P-Boat in the voyage down. Like someone violently wrenching a telephone from the wall, you cut us off from

---

[26] Friehe and Stommel, "Andrew F. Bunker."
[27] Simpson Papers, 1.6, Personal Diary I: 3–5, Oct. 16, 1952.

reaching each other…Your face distracts me and I cannot go on; emotions of the present forbid synthesizing those that have gone before (Figures 3.4 and 3.5).[28]

Back in the tight-knit community of Woods Hole, Joanne had to confront her divided loyalties. She had returned from the field campaigns "radiant" with intense happiness, turning over in her mind's eye every detail of time spent with Claude. After a candlelight dinner at Claude's house, she wrote in her diary, "I felt like a princess, in a dream…When I think of dancing from now on, it will be of dancing with you in the darkness, almost madly, till we were exhausted and dizzy and held on to each other to keep the room from spinning…Why is it that doing anything with you makes that same act under any other circumstance afterwards seem flat by contrast?" She counted the moments they shared as the greatest of her treasures: "Even the small fraction of

Figure 3.4 "Joanne Malkus working at Woods Hole 1956." Simpson Papers, 455, Photos. PD-9.

[28] Ibid, 17–19, Nov. 8, 1952.

Figure 3.5 Claude Ronne. "Picture of C in about 1955, age 45." Simpson Papers, 455, Photos. PD-13.

these that come quickly to mind in a few seconds are more wealth than I anticipated to find in this world."

> [We stood] in the cold calm, biting May dawn at quarter to five Monday morning… on the slight elevation behind your house, my arm in yours, with the whippoorwill informing us that the sky was reddening in a fringe above the dark wet trees. And it is spring again and it was dawn again, and we have been a long way since the last spring and the last dawn. And the Monday dawn was the more moving because that morning when the sun came up it ended a night we had shared—the chess game, Sully's arrival and departure, the emotional crescendo which began by my recounting my ballet dream.[29]

People were talking, however, and the summer of 1953 was not a good one. "The summer was a nightmare, really. At first on coming back from Puerto Rico I found it very hard to adjust to not being with you, to not sharing everyday life with you, every breakfast, every

---

[29] Ibid., 107–14, May 5, 1953.

Figure 3.6 Joanne, Willem, Richard Scorer, and an unidentified person, possibly Claude Ronne, dining at the Malkus's small house on Salt Pond Road, Falmouth, September 1953. Simpson Papers, 455, Photos. PD-12.

trivium, almost every thought. I probably found it harder because I could catch glimpses of you running around the Oceanographic; I could talk banalities with you in the office from time to time, and no more. And then came the gossip, which upset you so."[30] Joanne was distraught, and Willem had grown suspicious of Claude's frequent visits to their small house on Salt Pond Road in Falmouth (Figure 3.6).

By contrast, Joanne called the summer of 1954 a "wonderful one" with Claude: "I had so many glorious times with you and we shared so much and did so much together that time flew by too fast to cherish each moment in writing then and there." She found their conversations therapeutic, their moonlit outings memorable, and their lovemaking transformative:

> In early summer we waded barefoot up the Mashpee River [Cape Cod] and ate our picnic and drank daiquiris out of a thermos in a sunny clearing in the woods, and for some reason I was moved to pour out to you my early childhood traumas and jealousies—and you were so kind and understanding that many of the hurts seemed to go away. One late June

[30] Ibid., 134–42, Nov. 11, 1953.

sunset, we drove down to the Riverway in West Yarmouth, and after a discussion over manhattans (of what we admired most & valued most in a person, among other things) we took a walk in the full moonlight. Through the elm-arched streets we passed some of your old haunts and wound up down by Bass River. Marsh-smell, all full moon on breeze-flecked estuary—my arm in yours, how different and yet reminiscent of our late night walks around the base and docks in San Juan.

The summer got even better as it went on, from my unexpected arrival covered with blood and poison ivy (and you welcomed me and comforted me), to your birthday and my last Saturday night with you. On these last few occasions, our relation gained a new facet, a new depth and closeness. You opened up and talked to me in a way I never dreamed possible. You told me that our relation had changed your whole life, had affected all your relations with others. You made love to me. Oh, it was not too late; it was not, despite what either of us said before.[31]

Return to England

Joanne was determined to return to England to study with Scorer and the cloud physics group at Imperial College. She approached Columbus Iselin, who suggested applying for a Guggenheim Fellowship. Iselin had connections with important persons worldwide thanks to his reputation in oceanography, his high social position, and the Iselin family fortune. He was a personal friend of Henry Allen Moe, administrator of the John Simon Guggenheim Memorial Foundation, and wrote a letter of recommendation on Joanne's behalf in support of her fellowship application. Joanne was far from sanguine about her chances. While a student at Chicago, she had been turned down for a fellowship in Sweden, in part because she had been unable to get a good letter of recommendation from Rossby. She also suspected her chances were low to nil, perhaps one in ten, because she was a woman. Thinking that there was not much of a chance, she said to herself, "If I get this fellowship, I'll take ballet lessons."[32]

Ever since she since she witnessed her first performance of the Russian Ballet at about age seven, Joanne wanted to take lessons, but her mother never let her, saying they were too expensive. By the time she got into the higher grades in school she did not have time to do it

---

[31] Simpson Papers, 1.7. Personal Diary III: 295–7, Sept. 10, 1954.
[32] Simpson, "Oral History ," by Harper.

anyway. Ballet receded from her agenda, except for watching when her father's newspaper position provided free tickets. She was awarded a Guggenheim fellowship to study at Imperial College, and she did take ballet lessons. She claimed she had an absolutely marvelous time doing ballet as a serious hobby until about 1970.

Joanne took the entire family with her to London in August 1954 on the Cunard luxury liner RMS *Queen Elizabeth*. Willem was supported by ONR and worked in their offices in Grosvenor Square. The trip was exhilarating, but far from luxurious. They left Woods Hole in the midst of Hurricane Carol (Joanne said they "waded" to New York through the wreckage from the storm) and soon learned that traveling "tourist class" meant that their accommodations were in steerage, below decks, with cramped quarters, terrible coffee, and supercilious looks from the crew.

During the voyage, Joanne wrote a very long letter to Claude, copying its sixteen pages into her personal journal.[33] Its essence concerned their recent physical relationship and its therapeutic value for both of them. It delved into the psyche of each person, revealing details of Claude's narcissistic tendencies and Joanne's self-deprecating personal image. After several pages spent philosophizing about friendship, love, and sex, Joanne indicated that she and Willem had been discussing her affair, or perhaps hypothetical affairs in general. "Where would we have gone, what would our relation be like, had we not loved one another physically? For my part, I think we should have survived very well and continued in a close and growing friendship."[34] Willem thought that such a relation would be unnatural, would impose a great strain, and therefore would be unstable. Joanne agreed with Willem. She said she was "very very happy" that, over the past eight months, her relationship with Claude had indeed become a physical one. She wrote in her journal:

It is so much easier to talk freely to a person, to sense what he feels, to interpret his words with love and understanding, when you are laying with your rumpled head against his chest and can hear his every heartbeat, listen to every intake of breath, sense every hesitation and lifting of an eyebrow, when he has just demonstrated that he loves and trusts you; when the very situation is symbolic of dissolution of masks and

[33] Simpson Papers, 1.7. Personal Diary III: 298–313, Sept. 4, 1954.
[34] Ibid., 301, Sept. 4, 1954.

barriers—than when he is sitting in a nice white shirt and necktie in a chair six feet away—some of the habitual shell we adults have spent years in depositing around our souls remains, even if we are best friends.[35]

Willem was angry and abusive. He admitted he was having affairs, but Joanne said she was secure and confident that he loved her and had "enough love left over to share with others."[36] She rationalized her own affair with Claude by claiming she was paying forward a gift from Willem that more than compensated for his anger and physical abuse. She wrote about this to Claude:

> For the first time in your life a woman (whose opinion you can respect) has convinced you by word and deed that she really loves you—not some image, mask, or pose, but really you. Therefore the notion that you have carried around so long, that you are an unlovable outcast, that no one would love you, has had to go away—because [of] one genuine exception, one real person whom you cannot explain away as deluded or psychotic—that breaks the jinx. I understand this very well having had such a very similar experience myself and in my case it was Willy who did this for me—I could not in my great respect for him and his character and judgment explain away the fact that he loved me by saying that he must be crazy, because he obviously wasn't. Therefore this meant that I could not have been such an utterly hopeless and unlovable wreck myself after all—and so the jinx was broken, and it never came back in full force again. And I knew I could never repay Willy for this—one can hardly ever repay a good turn to the same person who has rendered it, but if I have passed the same sort of thing on to you to any extent at all, then I feel the more deserving of it in the first place.[37]

There was one big problem: she was still married to Willem—and the children (nowhere in her entire archive does she say much about the children) were literally in the same boat, by her side. She turned to her journal, attempting to analyze the problem rationally, as if she were taking on a new degree program or examining a scientific puzzle. She wrote in a letter to Claude, copied in her journal: "On the course we have chosen now the difficulties lie probably more in coping wisely with externals than with internals. It is an unconventional situation, a compromise situation (but that is life) and one requiring maturity and self-discipline—both of which you have exhibited to a high degree. But

[35] Simpson Papers, 1.7. Personal Diary III: 302, Sept. 4, 1954.
[36] Simpson Papers, 1.6, Personal Diary II: 268, April 23–25, 1954.
[37] Simpson Papers, 1.7. Personal Diary III: 310–11, Sept. 4, 1954.

speaking for myself, I usually prefer to navigate the more difficult and challenging course—nothing in life that is worth achieving is gained without compromise, without self-discipline, or without effort and some or many heartaches along the way."[38]

Joanne arrived in Southampton emotionally unsettled, faced with the tasks of moving the family into their unheated flat in West Kensington and establishing her presence at Imperial College, where she worked on a numerical cloud model and observed analog experiments that demonstrated convection (see Chapter 4). Although she was in London with Willem and her two boys, clearly her heart was with Claude, and she wrote long love letters to him:

> It is now ten months since I have written anything in these notebooks. Somehow when I wrote you that long, long letter from the Queen Elizabeth on the way across, I thought that perhaps I would not feel the need to write in here again, for everything I had to say, everything I had to report or comment on could be so easily said to you. Without the wall which I often sensed between us at the beginning of our relation, what need would there be to write here? And so it was, while I was away from you. The writing went to you, sealed in envelopes, instead of in here. Incredibly long letters, some of them love letters (I'll admit it now) and more showing the love more subtly, by a need to keep in touch with you, keep close to you, to share my experiences and reaction with you.

As she toured the historical sites of England, Joanne longed wistfully for Claude to be able to share her experiences. She visited Tintern Abbey and imagined Claude was by her side, viewing the beautiful green hills and trees framed in the ruined grey gothic arches; She went to Gloucester Cathedral and wished Claude too could see the delicate towers by moonlight, the sun through the stained glass of the lovely Lady Chapel, the angel choir. She wished he could feel this and see it and hear the bells ringing on a Sunday morning.

> In London, I walked around the streets and Georgian crescents, rode on red upper-decked buses, rushed with crowds through the windy, many-lighted corridors of Charing Cross Station, hunted for pre-fire churches and burying grounds near Cheapside, in the old city. Your sensitivity and powers of observation would have encompassed this, and enriched it for both of us. I tried to develop some of these faculties myself and to describe the scenes to you ... and the events of everyday life, the amusing incidents, work, shopping, afternoon tea drinking at the College—the

---

[38] Ibid., 304, Sept. 4, 1954.

people, their intricate characters and complex relations between them... All this I wanted to share with you, hear your comments on; the letters, long as they were, could only be so partial, so fragmentary.[39]

Her dream came true when Claude visited the project for six weeks:

Yes, now you have been here, actually here in England with me for six weeks, and the happiness I feel tells me this was no dream. And you have been to Tintern Abbey with me and we stood together on the stone steps in the north transept and looked across the ruined choir, floored with its incredibly green soft grass, and out the great south window where the few remaining delicate arches and circles framed a sheep-covered hill, a large oak with a black trunk, and a dark grey British sky. We saw Gloucester Cathedral and the cobbled courtyard of the New Inn, with its vines and potted geraniums. We climbed the ruined tower of Raglan Castle in the clear sunshine and shadow of a cold June afternoon and picnicked above the banks of Wordsworth's sylvan Wye...[40]

The background was their daily life in London: "It was wonderful having you work in the same room with me—to hurry in on the Underground in the mornings and find you. If the office door was ajar as I went to push it open, my heart would lift because I knew you'd be there, waiting to share the morning cup of coffee with me. We would sit near your desk discussing things, abbeys and strikes, politics, work, people, until Scorer busted in, in a flap to get on with the experiments."[41]

Joanne remembered, fondly, visiting the London Museum with Claude to show him the models of old Saint Paul's. She found it "wonderful" to walk through Kensington Gardens with him on a hot summer afternoon, with a new-cut grass smell and trees in rows, floating their feathery catkins, and to drop exhausted under one, on some rickety old chairs, to smoke a cigarette. She told him it was "wonderful" to walk together in St. James's Park in the late evening and to show him the exact spot where she had imagined, almost believed, the hallucination that Claude was with her three-and-a-half years earlier in the January dusk. She was touched when Claude told her that he had started to go back there alone and found that he could not do so without her. Joanne counted it a "sure revelation" that Claude felt the same way she did, adding, 'How many places in London now shall I hesitate to return

[39] Ibid., 314–22, July 1–2, 1955.
[40] Ibid.
[41] Ibid.

to?"[42] But most of all, and perhaps best of all, Joanne recalled the many, many times she hurried off at the end of her ballet class, rushing for the quickest shortcuts on the Underground, and arriving in the leafy late afternoons at Claude's garret in Onslow Garden, "where you were waiting for me, and had a drink and supper waiting for me, and we had the whole evening before us to talk and be together."[43] Here, with the green treetops outside the windows and the chimes of St. Paul's and its notched spire just barely visible through the concrete balustrade, the close relationship between Joanne and Claude grew and developed, like a hot tower cumulus cloud over the tropical ocean.

Her heart was overflowing as Joanne recalled seeing Claude off at the airport. As they sat on the train and in the dark-beamed typically English tavern waiting for the taxi, she tried to memorize every line of his beautiful face...

> Your profile which I admired so often outlined against the blister of the PBY, against the trade-wind clouds and the harbor of San Juan, I shall now remember forever as it was the other evening, framed against my bare shoulder—your eyes half shut and breath coming swiftly in a moment of passion and love, soon to be realized so gloriously. Your mouth was drawn back in the corners in a thin, tight line, almost as if something were hurting you, and your cheekbones and eye sockets looked as if a divine sculptor had chiseled them. What a contrast with a few moments later when you lay relaxed in the beginning dusk with your head slightly tilted back on the pillow. Your lips looked fuller, the contours of your face softer in complete repose—no wonder then that the rustling of the tree tops made me say that watching you so contained more beauty than all the trees in Kew." As Joanne stood alone in the airport terminal after seeing him off, "with your kiss still moist on my mouth and the lovely comforting smell of your sweat still faintly on my shoulders and body, I thought it only a matter of moments until the shock would wear off and I'd begin to cry my eyes out. But it didn't happen that way. Too much has been gained, too much security and strength in this relation to bemoan a temporary separation; too much happiness in the last six weeks to fade so soon."[44]

Joanne cherished her relationship with Claude and the record she kept of it. She read her notebooks over and over again, allowing them to trigger her memories, both sweet and bittersweet. Although she lost

[42] Ibid.
[43] Ibid.
[44] Ibid.

all of her personal photograph albums when she divorced Willem, she kept the three notebooks, the notes, and the letters with her, "so that I shall have it many years from now to recall some of the best moments life brought me."

> The other night I read over these notebooks – from start to finish, late into the night. I couldn't stop reading. In a way, it was as if someone else had written them; I found myself several times choking up and nearly in tears. And yet it was obvious all the while to me that I had written them, because each page brought back so many, many associations, and incidents that had not been recorded, but which were recalled by these few that were. Especially incidents from our trips to San Juan and St Thomas, which I would not lose from my memory for all the honors this world could offer. I think if my house or office should burn down and I only had time to get one thing out, it would be these notebooks (provided, of course, that all humans were safe and sound). I think I would rather let all the notebooks full of equations burn up far sooner, though I am probably a better formulator of equations than of emotions.[45]

Joanne's self-analytical style and romantic inclinations are in full display here. She did not place restrictions on her archived journals. It was part of her desire to be understood beyond her professional résumé and to unmask her emotions.

As she had promised herself, Joanne sought ballet lessons during her research sojourn. She called all the best schools in London; all but one flatly rejected her. The exception, in Hampstead Heath about an hour's commute away, let her try out for grade 1 with 7- and 8-year old girls (Joanne was 31). She went regularly three times a week, advancing to grade 4 with 12- and 13-year-olds, before she left England for Sweden, where she continued her lessons without a working knowledge of Swedish. She later studied at the Cape Cod Conservatory with Jane Gay Stephens, who had been a soloist with the Russian Ballet. Joanne persevered, towering over the younger girls in the chorus, or taking adult parts like the evil old witch in "Sleeping Beauty" or the genie in "Aladdin and His Magic Lamp," which she considered her best solo part and one of the high points of her life. She wore a gold costume and turban and had a fairly difficult toe part in performing her magic. Such performances came to an end in 1961, as no ballet school in Los Angeles would accept her, but she got back to it in 1967 in Miami, where

---

[45] Ibid., 286–87, 297, May 8, 1954.

Figure 3.7 Joanne Malkus in ballet costume in the front room of her Falmouth house, 1959. Joanne's caption: "The picture of me was taken by the Woods Hole photographer and friend, Claude Ronne." Simpson Papers, 455, Photos. PD-3. Original in color.

she studied with a top teacher and learned to do pair work on toe. She persevered for several years until the demands of the school no longer fit her schedule (Figure 3.7). She had started too late, but because she loved it so much she put her pointe ballet shoes, which meant so much to her, in the 1985 Chicago Museum of Science and Industry exhibit, "My Daughter the Scientist," which traveled to other science museums for the next decade (Chapter 8). The ballet shoes are in the Radcliffe collection.[46]

Pacific Cloud Hunt

In the summer of 1957, Claude and Joanne hitched rides on military cargo planes shuttling across the Pacific. Using calibrated time-lapse

[46] Simpson Papers, 19F+B.2m. Exhibit item: Simpson's pointe ballet shoes used in display representing Simpson's work and outside interests, in "Women in Science" (later renamed "My Daughter the Scientist") traveling exhibit (1985–1995).

cameras, they photographed clouds and collected synoptic data during transoceanic flights in what they called the "Pacific Cloud Hunt." Joanne wrote a first-person account of their meteorological "hitch-hiking" adventure for the University of Chicago Magazine.[47] She described how she, a female meteorologist, and her associate Claude Ronne, climbed aboard an aging Douglas Globemaster loaded with 30,000 pounds of diverse cargo for a noisy, bumpy, seemingly interminable flight from Hawaii to Manila and Bangkok, with refueling stops at Wake Island and Guam. "Why do we do this?" she asked. "What strange concatenation of circumstances brought us there, huddled between films, cameras, notebooks, and snoring airmen … in a flying barn eight thousand feet above a ship-less ocean?" The scientific goal was to photograph and map common (small cloud) and rare (monster cloud) formations as they influence tropical weather systems. Her unstated personal motivation, however, was to be with Claude on a romantic quest. First things first: could a woman get the necessary permission and military orders to board a cargo flight? Joanne showed that it was indeed possible. The pair of Woods Hole researchers crisscrossed the tropical Pacific on Military Airlift Command flights at heights of up to 9,000 feet. Over the course of about 80 hours, they exposed some 8,000 feet of film at 1-second intervals (Figure 3.8).[48]

Just after taking off from Hawaii, the pilot, Captain Rodgers, invited Joanne to sit in the co-pilot's seat in the cockpit. She was awe-struck by the view. Ahead and below her spread a panorama of clouds, glaring white against the blue Pacific. The glittering towers of the clouds looked like hard snowy mountain crags as they approached, but dissolved into grey jouncy fog, splashing the windshield and leaking rainwater into the plane. Here and there a tall icy anvil fanned out skyward. Claude wrote in his notebook, "Clouds this morning are high and exultant."[49] Joanne felt this way too.

Cumulus cloud research by aircraft and by photography had long been part of the Woods Hole meteorology program. Cottony, cauliflower-headed clouds grow most abundantly over the tropical oceans and are often seen as decorative backgrounds in postcards along with palm trees and coral atolls. These trade-wind cumulus clouds appear over

[47] Malkus, "Pacific Cloud Hunt," 4–10, 24.
[48] Simpson Papers, MP35.1. "WHOI–Pacific Clouds–[Claude] Ronne & [Joanne] Malkus (Simpson): Clouds in hurricane breeding region of North Pacific between Marshall Islands and Mariana Islands … Flight III." August 16–18, 1957. 16-mm film, 14 minutes.
[49] Ibid.

First Woman

Figure 3.8 Joanne with movie camera, Pacific Cloud Hunt, 1957. Simpson Papers, 455, Photos. PD-9.

the oceans in undisturbed weather conditions and are remarkably uniform in size and shape. Although abundant, their vertical development is suppressed by the dry, stable air above them—the so-called trade wind inversion (Figure 3.9).[50]

In certain areas, where the air is unstable, clusters of moist updrafts pump water vapor fuel from the warm ocean surface to higher levels, forming much larger clouds and cloud clusters. These monster clouds, with tops reaching 30,000 feet or higher, are rare indeed. They produce rain and release heat, ultimately driving the winds, storms, and general circulation of the atmosphere. The Pacific Cloud Hunt focused on these monsters (Figure 3.10).

In August 1957, as their airplane skimmed the edge of super-typhoon Agnes, Joanne and Claude felt they had caught a glimpse of the breeding ground of storms. Perhaps their film would reveal how the huge monster thunder clouds line up in rows, or rain bands, and constitute the "working

[50] Malkus, "Pacific Cloud Hunt," 8.

Figure 3.9 Tropical trade cumulus clouds, Simpson Papers, 455, Photos. PD-9.

Figure 3.10 Huge clouds in the Pacific Inter-tropical Convergence Zone at ~15 km. Lots of rain. These huge cumulonimbus clouds were in a line. Simpson Papers, 455, Photos. PD-9.

cylinders" of a giant natural-engine hurricane.[51] Perhaps their data could be used to improve computer simulations of clouds and help them understand why clouds and hurricanes are found where they are and in the patterns they display.

As the Pacific Cloud Hunt was coming to an end, Joanne recalled coming down the steps of the Pan American Clipper in Hawaii and being handed a telegram by Claude notifying her that Rossby had died of a heart attack while giving a seminar. Joanne was "completely cracked up" by the news and could not discuss his death for the remainder of the expedition. She remembered his remark to her, made just two months earlier, that she was really becoming a good meteorologist after all.

Meteorology was never the same again. Rossby was the last Leonardo Da Vinci of the atmosphere—that is, he was the last human who could and did know everything (or nearly everything) that was worth knowing in the field. His knowledge and creativity were capacious also in relation to the oceans. The many subdisciplines of atmospheric science had begun to emerge near the end of his lifetime. In Joanne's opinion, it was becoming so fragmented that specialists in cloud ice had no idea about blocking high-pressure cells, and vice versa.[52]

Joanne was featured on television in a 1958 CBS Conquest Program in a 15-minute segment in which she explained the reason for the Pacific Cloud Hunt, showed the cloud movies, and demonstrated how she had analyzed them (Figure 3.11).[53] In a video interview conducted in 2017, Peggy LeMone recounted how seeing Joanne on television in the 1950s inspired her to become a meteorologist.

> I learned only fairly recently that a film that influenced me strongly as a child had Joanne Simpson in it. It would've been somewhere around the mid to late 1950s—and I was in my brother's bedroom...watching his TV, and on came this fascinating piece about clouds. There was a woman looking at steam coming up from a cup of coffee and talking about clouds, and there were pictures of clouds and airplanes flying through clouds, and she was talking about how she studied clouds—and I was just totally delighted! Because, from the time when I was about eight years old and the lightning struck our house, I was fascinated by weather. I didn't know if girls could be meteorologists...but here she was, and she

[51] Ibid., 24.
[52] Simpson Papers, 3.9. Narrative (by Simpson): "at Woods Hole Oceanographic Institution. Summer 1947–1950" and "Pacific Cloud Hunt."
[53] [J. S. Malkus], "Origin of Weather," CBS Television Network, 16-mm film, sound, black and white, 26 minutes, 1960., YouTube.

Figure 3.11 Joanne on CBS Conquest program 1958, demonstrating how she projected and gridded Pacific cloud films in order to construct the maps in the 1964 book. Simpson Papers, 455, Photos. PD-14.

seemed to be quite coherent and seemed to know what she was doing, and I think that made me far more comfortable with my decision to become a meteorologist. So by the time I was thirteen years old, in seventh grade, and they asked, 'What do you want to be when you grow up?' I wrote: meteorologist. I think I might've had to ask the teacher to help me spell it. But! That desire continued, and that's what I am today.[54]

For the first decade of her career, Joanne kept a hand-written record of her research thoughts, derivations, and calculations—a discipline she considered essential in developing her ideas to the point of useful results, or more often to a dead end or blind alley. The notebooks prevented her from going down the same blind alleys too many times. More importantly, some of the apparent blind alleys much later blossomed into very important work after experience, discussions, or new observations gave her a totally different insight into the problem. Over and over again, the regular habit of recording her thoughts and calculations helped her develop new insights.

---

[54] LeMone, "Peggy LeMone on Joanne Simpson and CBS Conquest Program," YouTube.

She considered this "a useful lesson that young scientists or science histor-ians might find of value."[55] The set of research notebooks dated 1947 to 1957 document, in her estimation, about three-quarters of her most original work during this time as a largely unrecognized junior scientist.[56] In 1957 she ended this practice. Her last research notebook, labeled BESK, recounts her first attempt to simulate the growth of a cloud on a digital computer. With the coming of computers, she writes, "everything changed" and many meteorologists "became devoured by them." She found that the pro-cess of documenting and tabulating computer runs, both failures and suc-cesses, lacked meaning, and she quickly fell behind in the task of keeping a log of them. Agency and media had changed: "computers did the detailed steps of solving the equations, which were written out in those days on paper tapes or punched cards." Joanne lived through and contributed to a major shift in research methodology involving computer simulation. This involved abandoning the older practice of seeking linear solutions or approximations to complex phenomena, while modeling nonlinear feed-backs. She also thought that cloud modeling, rather than the more widely studied topic of numerical weather prediction, would be of interest to historians of computing.

Joanne was "breaking through"—working diligently to attain the professional recognition she had long sought, but her life was entering a new period of extreme emotional and professional turbulence. Although there were hopeful signs, discrimination against women was still common. In 1957, Joanne co-taught an oceanography course at MIT with her Woods Hole colleague, William Von Arx. She worked diligently on this course and prepared a set of carefully organized and typed lecture notes and problems superior to any available. Her MIT teaching came to a sudden and crushing end, however, mid-semester, when the department chair of geology and geography (not meteorology) discovered she was a woman. She was fired immediately and was not paid for any of her teach-ing or even reimbursed for the non-negligible commuting costs. Afterward, Von Arx was too embarrassed to discuss the fiasco with her, and Joanne kept waiting for a check that never arrived.[57] Unfortunately, this discrimination was, at the time, completely legal.

[55] Simpson Papers, 9.2. Narrative by Simpson, January 1994; re: description of under-graduate and graduate college career.

[56] Simpson Papers, 9.3. Work Notebook #1, development of PhD dissertation, August 1947–June 1949.

[57] Simpson Papers, 6.7. Teaching: Simpson's narrative re: teaching at UCLA (and elsewhere), 1993.

4

The Path to Hot Towers

Hot towers are the driving force in the tropical atmosphere, providing energy to power hurricanes, the Hadley circulation, the trade winds, and by implication, the global circulation of the atmosphere.

RIEHL AND MALKUS[1]

Tropical Meteorology

When Joanne started her career, first as a graduate student at Chicago and then as a research scientist at Woods Hole, the field of tropical meteorology was undergoing an era of vigorous growth.[2] Meteorologists lacked the capacity to issue reliable forecasts in the tropics, even though they cover half of the world's surface and are home to about half of the world's peoples. Although it was known to be a region of great importance, there was no scientific agreement on the important ways the tropical atmosphere influences the large-scale behavior of the globe. It was known, however, that transfers of energy occur there through processes such as absorption of radiation, evaporation and precipitation, convection, and hurricanes. Herb Riehl at Chicago had just instituted the first courses on the subject, and Joanne was his most eager student.

The venerable field of tropical meteorology—tracing its heritage, in the age of sail, to Edmund Halley's global wind chart (1686), the sun-driven direct circulation of George Hadley (1735) and the tropical storm theories of William Reid and William Redfield (1840s) had been overshadowed by studies in northern temperate and polar latitudes, as exemplified by the familiar polar front theory of the Bergen School in Norway. World War II in the Pacific had increased the amount of available observations and had exposed many meteorologists to the scale and complexity of the

---

[1] Riehl and Malkus, "On the Heat Balance in the Equatorial Trough Zone" (two papers); Riehl and Simpson, "Heat Balance of the Equatorial Trough Zone, Revisited."
[2] DeMaria, "History of Hurricane Forecasting."

First Woman: Joanne Simpson and the Tropical Atmosphere. James Rodger Fleming,
Oxford University Press (2020). © James Rodger Fleming.
DOI: 10.1093/oso/9780198862734.001.0001

challenges. Post-war nuclear testing in the Marshall Islands also provided new data and generated urgent needs for accurate forecasting.
Although Harry Wexler had recently used the new technology of radar
to study a hurricane, there was, as yet, little observational capacity, no
predictive or dynamic models, and no generalized theory of the tropical
atmosphere.[3]

In 1951, the AMS published a massive 1,334-page *Compendium of
Meteorology*, a state-of-the-art review of all aspects of meteorological
theory and practice. "Tropical Meteorology" comprised four articles
occupying just fifty-four of these pages.[4] With the exception of
A. Grimes, who worked in Malaysia and West Africa, the authors—
Clarence Palmer, Gordon Dunn, and Herbert Riehl—were all alumni of
the Institute of Tropical Meteorology, established by Rossby in 1943 in
Puerto Rico. Palmer was the Institute's first director, Dunn a hurricane
expert, and Riehl, a University of Chicago professor and Joanne's PhD
supervisor. Although Joanne's work was not represented in the volume,
it is likely that she had some input into Riehl's article; and it is certain
she would have read all the articles very, very carefully.

Three distinct schools of thought dominated: the climatological, the
airmass, and the perturbational. The climatological approach emphasized monthly, seasonal, and annual means of features such as the trade
winds, monsoons, and the doldrums. Adherents of this approach focused
on statistical analysis, mean maps, and large-scale motions. They were
convinced that the day-to-day weather in the tropical Pacific differs very
little from that revealed by monthly and annual means. They extended
this approach to the forecasting problem. Some felt, for example, that
the best guide to forecasting the tracks of individual typhoons and tropical storms might lie in the study of mean storm tracks. Their dynamic
theories, when they had any, were expressed in terms of a hypothetical
general circulation of the tropics, and this was the entity whose characteristics were to be explained, not by the vagaries of the daily weather
map, but by physical reasoning from first principles. This can be called
the Climatological School.

A second group, inspired by the Bergen School, attempted to import
concepts from high-latitude airmass analysis into the tropics. Some

---

[3] Wexler, "Structure of Hurricanes."

[4] Palmer, "Tropical Meteorology," pp. 859–80; Grimes, "Equatorial Meteorology,"
pp. 881–6; Dunn, "Tropical Cyclones," pp. 887–901; Riehl, "Aerology of Tropical Storms,"
pp. 902–13.

argued that the tropical atmosphere differed from that in higher latitudes only in having a higher temperature and an easterly instead of a westerly mean wind direction. They looked for air masses and an "equatorial front" similar in principle to the polar front, and occlusion processes they assumed led to the formation of typhoons and tropical storms. Small temperature differences were important to this group, and the "slope" of systems became a matter of earnest discussion.

A third group—the Chicago School influenced by Rossby—adopted the statistical results of the climatological school, adding that the basic currents of the tropical atmosphere were subject to upper-air perturbations implicated in surface weather conditions. Although all of the traditional approaches to studying the tropical atmosphere had their deficiencies, Rossby's perturbation method, combined with more complete observations, especially of the upper air, seemed to hold the most promise, since the tropical atmosphere, more closely than any other region on Earth, approximates the motions of an ideal fluid. In the tropics, temperature and pressure differences are typically weak (with the notable exception of tropical storms). This school of meteorologists learned to base their analyses and forecasting on models of wave-like perturbations propagating through the wind field.[5] Dunn and Riehl promoted the theory of "easterly waves" in the Atlantic that regularly form into organized bands of intense convection and, on occasion, develop into tropical storms.[6]

It had long been known that tropical cyclones are the most violent weather phenomena on the planet. In his article in the *Compendium*, Riehl reviewed the current knowledge on the structure, formation, and movement of tropical storms, shortcomings of this knowledge, and suggested paths for future research. He pointed to the lack of observations, which was so keenly felt in all low-latitude work, and called for a network of upper-air data observations, especially during the hurricane season. He addressed, head-on, the issue of whether tropical cyclones were shallow phenomena, extending no more than 3 km in height and thus controlled by the sea-level and lower atmospheric conditions, or whether they extended to great heights through the depth of the troposphere. He cited the theoretical argument of Bernard Haurwitz and recent high-level observations that indicated that mature

[5]   Palmer, "Reviews of Modern Meteorology."

[6]   Dunn, "Cyclogenesis in the Tropical Atlantic"; Fleming, *Inventing Atmospheric Science*, pp. 107–8.

storms extend through the troposphere.[7] Riehl was convinced that perturbations in the upper-air flow were the keys to understanding them.
The limitations of the tropical observing network led him to conclude
that statistical/climatological approaches to hurricane behaviors were
currently the best available strategies, at least until more flights, organized field campaigns, and possible satellite coverage became available.
Riehl's pioneering course in tropical meteorology, much of it incorporated into his landmark textbook, *Tropical Meteorology* (1954), shaped
Joanne's professional entry into a field that was wide open.

Toward a Theory of Convection

During her fellowship year at Imperial College in 1954, Joanne worked
with Richard Scorer on a numerical cloud model, which they com-
puted by hand. They were also conducting laboratory experiments to
simulate cumulus clouds in which a dense colored fluid (mud slurry)
was released into a tank of water to see if and how it entrained the
surrounding particles. These experiments convinced Scorer and
Ludlam that entrainment was plausible and informed Joanne's mathematical model of a growing cloud or "thermal." Although she was
there primarily to conduct research, Joanne unwittingly volunteered
to do a little teaching. The head of the department, Professor P. A.
Sheppard, immediately assigned her the synoptic laboratory—the
one course she did not feel qualified to teach, since she knew nothing
about European weather or the British forecasting system. Joanne
was horrified, and had to spend a lot of time preparing lessons. She
learned from the staff of the Meteorological Office how to forecast
and how to instruct students using a little wall chart—a weather
map centered on Great Britain that excluded Europe and most of
the ocean.

When Rossby visited Imperial College that year, Joanne showed him
her numerical model. It was a table-sized grid of equations that took her
three months of slide-rule calculations to compute only a few time-
steps. The paper grid had erasure holes all over it. Rossby remarked,
"You'll never get anywhere with this. I have one of the brand new computers that's just like the ENIAC [used by the Meteorology Project] at
Princeton. Why don't you come [to Sweden] and I might be able to get

---

[7] Haurwitz, "Height of Tropical Cyclones."

somebody to help you with the programming."[8] After examining their tank model, Rossby remarked to Scorer, "Well, you're just artificially making this bubble and then observing it. How do you know that nature makes the bubbles?" Scorer replied, "We have sampled . . . clouds with gliders and we're pretty sure that the bubbles are there." Rossby, who had extensive experience interpreting observations and setting up rotating dishpan experiments, was unconvinced and wanted to see if the numerical model could simulate a vertical plume like that seen in the tank.[9]

Joanne's work was a fundamental attempt to understand thermal convection. It was a nonlinear problem best modeled on a digital computer. Nature had to concentrate weak temperature gradients into strong ones to produce a rising thermal with an internal vortical (or rotating) circulation. Starting with a slight buoyancy, the bubble of air starts to rise and causes motions in the air and near it. The movement of the air changes the temperature distribution, and the changed temperature distribution modifies the buoyancy and thereby the motion field. The temperature and motion are so closely interactive that changing one field changes the other. Thus the problem is nonlinear.

At age 30, Joanne was beginning to receive the recognition she had worked so hard to achieve, though not yet from Rossby, who made the tactless comment that she was "sticking out like a sore thumb." Still, she believed that being a woman was an asset rather than an obstacle.[10] She had developed a strong working relationship with her mentor, Herb Riehl, and their research was complementary. Joanne focused on tropical cloud phenomena of mesoscale (~10 km), while Riehl specialized in synoptic-scale (~1,000 km) phenomena. Working together, they established a new view of the dynamics of the tropical atmosphere.[11] Years later, she told Riehl that she often referred to his 1954 tropical meteorology book and considered it a classic. She joked that other people come into her office and "steal it fairly often." Noting that a good book's lifetime is usually about ten years, she found it quite amazing that "there's nothing in it that's wrong . . . although we have learned

[8] Simpson, "Oral History," by Harper.

[9] Fleming, *Inventing Atmospheric Science*, pp. 81–2.

[10] Simpson Papers, 2.10. Clippings: "Beginning of research career in atmospheric sciences—Joanne Malkus Simpson, 1953–1964, early recognition, Imperial College, London-Guggenheim Fellowship-clippings."

[11] Riehl and Malkus, "On the Heat Balance and Maintenance of Circulation."

more things in more detail since then...And therefore you could still base a course in tropical meteorology on it, which in fact I have done, and just add other pieces and publications that are later to fill out some of the things."[12]

A Numerical Cloud Model

In 1955, Joanne took Rossby up on his offer to visit the International Meteorological Institute in Stockholm. She was in residence from August to November, while Willem worked at the ONR office there. Joanne attended lectures by Bo Döös and Arndt Eliassen on numerical weather prediction, and worked with Georg Witt, who programmed her cloud model into the BESK—at the time, the most powerful computer in the world.[13] By the end of November they had nearly finished a nine-minute run of a bubble development in which the original temperature perturbation was at the ground. One could see a vortical circulation developing and the temperature gradient at the bubble top becoming tighter as the bubble slowly rose and intensified. After a lot of false starts, truncation errors, and other difficulties in computation, the model was becoming interesting, and results were beginning to resemble something reasonable. Reluctantly, although she had not yet finished, Joanne and the family had to go back to Woods Hole, since she had already been away for nearly eighteen months.

Back at her desk in Woods Hole in January 1956, as she was thinking of improvements for the numerical model of a rising thermal, a letter from Rossby arrived, and she wrote in her research journal, "Back to the trade-wind problem—at Rossby's behest."[14] An informal seminar she had given on a model of the trade winds had motivated Rossby to ask her to write up the work for the Swedish geophysical journal *Tellus*.[15] She also renewed her collaboration with Riehl on the heat balance of the trade winds.[16] Joanne considered these papers to be two of her most important.[17]

[12] Riehl, "Interview," by J. Simpson.

[13] Ibid., 119.

[14] Simpson Papers, 9.3. Work Notebook #1, development of PhD dissertation, August 1947–June 1949.

[15] Malkus, "On the Maintenance of the Trade Winds."

[16] Riehl and Malkus, "On the Heat Balance and Maintenance of Circulation."

[17] Simpson Papers, 10.4. Work Notebook #6: England, Sweden and start of Pacific Investigation [which included more on the bubble theory of convection], 1955–1957.

To prepare for Rossby's anticipated sixtieth birthday celebration, the planners asked Joanne and Georg Witt if they would contribute a chapter on their cloud model to a *Festschrift* in his honor. Joanne responded, "Well, we could, but it isn't really quite finished." ONR sent Joanne back to Stockholm that June to finish the model. They had to work when the computer was available, typically midnight to 3:00 am or 3:00 am to 6:00 am On a Sunday night two days before her departure, as they had just about completed the model, they arrived to find the door to the computer room locked, and so Joanne called a locksmith to let them in so that they could finish the model runs. When Rossby found out about this, he did not know whether to be angry or to laugh. He finally wound up laughing, remarking, "Well, that's a pretty nervy thing to do." So they finished the paper—the first (crude) computer model of a convective element—and sent it in.[18] It pleased her to hear the great Rossby say to her on that occasion, "So you are making a big contribution after all, so everything it cost turned out to be worthwhile." She was grateful he said this, and that he had changed his attitude, especially since it turned out to be the last time they were together. Still, gender dynamics were in play: "He also began to flirt with me, innocently but rather over-obviously, at dinner parties. I was embarrassed because I wanted him to remain a father figure, and also his wife Harriet was one of my closest friends."[19]

The National Hurricane Research Project

In 1954, three devastating land-falling hurricanes, Carol, Edna, and Hazel, raised public outcry and convinced members of Congress that funding was needed for research on severe storms. Carol caused sixty deaths and $461 million in damages along a track across Long Island and New England in late August. Only eleven days later, Edna followed a remarkably similar path, resulting in twenty deaths and $42 million in damage. Hurricane Hazel provided the knockout punch, causing as many as 1,000 deaths in Haiti and devastating the island's economy. Hazel made its second landfall in the Carolinas in October as a Category 4 storm, then tracked northward through the mid-Atlantic states and

[18] Malkus and Witt, "Evolution of a Convective Element."
[19] Simpson Papers, 3.10. Narrative (by Simpson): "Malkus–Riehl Collaboration (1947–1961)."

into Canada. The storm resulted in almost 200 fatalities in the US and Canada and up to $300 million in damages.[20]

In December 1954, Joanne was invited to a high-level conference in Washington, convened by experts in tropical meteorology, and tasked with formulating a federal response to the devastating hurricane season just past. Here she worked with hurricane forecaster Bob Simpson, who recalled first meeting her in March 1943 when he presented a seminar in Chicago on his adventures in tropical meteorology. Their next encounter was at an AMS meeting in New York in 1952, when she spoke about cumulus clouds in the tropics. Bob recalled, in an understated way, "It was a presentation that captured my interest and admiration somewhat beyond the questions I asked about her work."[21] He was glad that Joanne had joined the hurricane advisory panel. Their recommendations led to an appropriation from Congress for a three-year National Hurricane Research Project (NHRP), to be managed by the weather bureau and directed by Bob Simpson. Initially, Herb Riehl was disappointed that the weather bureau did not ask him to direct the project.[22] He was the leading authority in the field, but had no experience negotiating the complex political landscape of Washington.

The NHRP was an all-out coordinated effort involving international and interagency partners to study the origin, structure, and dynamics of hurricanes, monitor their tracks, improve forecasts, and, if possible, seek ways to modify them.[23] It began in May 1956 and ran for thirty months, through the next three hurricane seasons. One of its overarching goals that went beyond individual hurricanes—a goal that caught Joanne's attention—was the study of large-scale energy transport in the tropics and from the tropics to other areas. There were many players: the United States, Great Britain, France, Cuba, the Netherlands, the Dominican Republic, Mexico, and Colombia. They supported a network of twenty-one rawinsonde stations measuring wind speed and direction, with each station launching two balloons each day. The US Air Force provided two weather reconnaissance B-50 aircraft, and the Air Force Cambridge Research Laboratory sent its instrumented B-47

[20] National Hurricane Center, "Hurricanes in History."

[21] R. H. Simpson, *Hurricane Pioneer,* p. 52.

[22] Riehl, "Interview," by J. Simpson.

[23] R. H. Simpson et al., "Objectives and Basic Design of the National Hurricane Research Project"; US National Oceanic and Atmospheric Administration. "Hurricane Research Division History"; Simpson Papers, 9.1. Writings: "The Riehl–Malkus Collaboration on Tropical Meteorology and Hurricanes," Memoir by Joanne Simpson, 1998.

jet. The aircraft flew in and around storms from near the surface to elevations above 12 km, and each carried a full array of advanced sensors, including Doppler radar, nose cameras, cloud droplet and particle sensors, and digital data recorders. The ONR proffered rocket launches to provide panoramic photographs of hurricanes and their environment from altitudes of 80 km or more.

By this time, Joanne had an established reputation in tropical meteorology, although she had not worked on hurricanes at all. In June 1956 she visited the project site in Palm Beach, Florida, for the first time. Bob Simpson welcomed her there, where they held "postmidnight sessions" discussing clouds and hurricane structure.[24] She was excited by the addition of punch cards to record data, which could be machine-sorted, with the printouts available within twenty-four hours, in time for their use in planning the next mission. Processing of manually collected or analog data often required months. She was impressed by the NHRP's level of funding and the quality of their instrumented aircraft, but she was not impressed by their policy that excluded women from the research flights.[25] She was quite taken by accounts of Bob Simpson's legendary flight into the eye of a Pacific typhoon in 1951,[26] as it was the kind of aerial adventure she admired. Flying out of Guam, Bob had spent three hours orbiting typhoon Marge, with its "gorgeous clear eye surrounded by boiling clouds like the ascending benches of a coliseum."[27] Joanne was getting interested in hurricanes—but then, hurricanes are just really big organized tropical cloud systems.

Hurricane Daisy

By late 1957, Joanne—now on the staff of the NHRP—enticed Herb Riehl to participate in hurricane research flights, the most significant of which were flights into Hurricane Daisy off the Florida coast on August 25–27, 1958. Legend has it that the NHRP injected silver iodide into this storm in an unauthorized, unreported, and unpublicized attempt to modify it. These were Herb's first hurricane flights and the first time he ever experienced measured winds of hurricane strength.[28] Although Joanne

---

[24] R. H. Simpson, *Hurricane Pioneer,* 98.

[25] Simpson, "Oral History," by Harper.

[26] R. H. Simpson, "Exploring the Eye of Typhoon Marge 1951."

[27] Simpson Papers, 9.1. Writings: "The Riehl–Malkus Collaboration on Tropical Meteorology and Hurricanes," Memoir by Joanne Simpson, 1998.

[28] Riehl, " Interview," by J. Simpson.

wanted to fly too, the military said "Sorry, no women are allowed to go."[29] It angered her that she had to stay on the ground and work with photographs and other information from the flights.

Joanne and Claude, with the assistance of Woods Hole technical assistant Margaret Chaffee, laboriously calibrated these data and turned them into maps which they then superimposed on the radar analyses.[30] The observations revealed a remarkable persistence of recognizable cloud patterns throughout the three days studied, and confirmed their hypothesis regarding the concentration of convective activity into a few lines of huge cumulonimbus clouds that occupied only about 4% of the rain area. Later, Herb and Joanne prepared a complete mass and energy budget study of the storm on its formation and mature days. They calculated the inflows and outflows of heat energy, angular momentum and potential vorticity (local rotation) in radial cylinders from the center outward at eight vertical levels.[31] The data allowed them to develop a model of the inflow layer for a steady-state hurricane based on the influence of "extra" oceanic heat brought into play when air spirals toward low pressure.[32] Although there was no satellite coverage, they were able to construct maps of the cloud formations by analyzing radar images and motion pictures taken by the aircraft. This rare encounter with a mature storm reinvigorated their research program and led to an ongoing question: "Hurricanes: Why are there so few?" (Figure 4.1).[33]

Hot Towers

At the First National Conference on Hurricanes in West Palm Beach, Florida, in 1958, Joanne proposed her "hot tower" hypothesis for mature hurricanes. She postulated that cumulonimbus clouds are the driving force in the tropical atmosphere, providing energy to power hurricanes, the trade winds, and, by implication, the global circulation. Her work revolved around clouds, especially tropical clouds. The little trade cumuli, extremely numerous but suppressed by the trade-wind inversion from ever growing too large (Figure 3.9), are metaphors for the many women in meteorology in the twentieth century, who remained in low-level jobs and whose careers were suppressed by discrimination from and

[29] Simpson, "Interview," by LeMone; Simpson, "Oral History," by Harper.
[30] Malkus et al., "Cloud Patterns in Hurricane Daisy."
[31] Malkus and Riehl, "Some Aspects of Hurricane Daisy, 1958."
[32] Malkus and Riehl, "On the Dynamics and Energy Transformations."
[33] Malkus, "Origin of Hurricanes"; Malkus, "Tropical Weather Disturbances."

Figure 4.1 Hurricanes: Why are there so few? Simpson Papers, 9.1. Writings: "The Riehl-Malkus Collaboration on Tropical Meteorology and Hurricanes," Memoir by Joanne Simpson, 1998.

expectations of a male-dominated society and profession. Yet, as Joanne showed scientifically, even the strongest trade-wind inversion could break down—a process that eventually occurred for women in meteorology, but only after Joanne had shown the way. The hot towers of the tropics (Figure 3.10) symbolize the heights that Joanne's career reached, soaring above all the rest; powerful and exceptional, yes, but attaining their pinnacle through extreme struggles against forces that would drag them down.

In formulating her hot-tower hypothesis, Joanne used data from their flights into Hurricane Daisy. She found that the heat released from just a few hot towers was able to sustain the warm core of the storm.[34] As she recalled in an interview: "All of a sudden I had a light bulb go on over my head. All the work I had done on clouds suddenly fit together with what a hurricane was doing, and I suddenly had an idea of how the hurricane heat engine works."[35]

By focusing on new observations and using them in a dynamic model, Joanne was able to solve an old conundrum involving the hurricane's eye. Traditional explanations of the calm central eye of a hurricane invoked the intrusion of dry air from the stratosphere. However, with the increased availability of radiosonde and aircraft soundings providing vertical profiles of the temperature and humidity, Joanne discovered that the relative humidity and the amount of water present were much too high for the air inside the eye to be undiluted air from the stratosphere. Further evidence for this came from a 1954 movie made by MIT of radar images of Hurricane Edna as the storm passed Cape Cod. The movie showed many pieces or fragments of the very tall eye wall apparently being drawn into the eye at high levels. Joanne took note of this and hypothesized that most of the air in the hurricane eye must have come from the cloudy eye wall. She was able to explain quantitatively how the forced descent of air produced a cloudless eye with light and variable winds at the surface.[36]

Once Riehl and Malkus recognized the importance of entrainment, they calculated that the core of giant cumulonimbus clouds had to consist of undiluted air. As Riehl recalled, "We christened these clouds 'hot towers'."[37] They were the driving force in the tropical atmosphere, providing energy to power hurricanes, the Hadley circulation, the trade winds, and the larger circulation of winds and storms around the Earth.[38]

---

[34] Simpson Papers, 3.10. Narrative (by Simpson): "Malkus–Riehl Collaboration (1947–1961)."

[35] Simpson, "Interview," by LeMone.

[36] Malkus, "On the Structure and Maintenance."

[37] Riehl, "Interview," by J. Simpson.

[38] Riehl and Malkus, "On the Heat Balance in the Equatorial Trough Zone"; Riehl and Simpson, "Heat Balance of the Equatorial Trough Zone, Revisited."

Only a few thousand of these giant clouds were sufficient to drive the planetary circulation, hurricanes included. "The furious winds [in hurricanes] are driven by latent heat released by giant clouds; their concentrated and highly buoyant character is also apparently essential in the heat and mass transports of the storm. The models envisage that the ascent in the core occurs in extremely restricted regions of rapidly rising cumulonimbus towers, which come to occupy a significant fraction of the rain area as the eye wall is approached. This latter concept has been informally christened the 'hot tower hypothesis'."[39]

It took Malkus and Riehl seven years to reduce the data collected on the Pacific Cloud Hunt. The resulting book, *Cloud Structure and Distributions over the Tropical Pacific Ocean* (1964), was devoted to a systematic analysis of each flight.[40] They included synoptic charts of streamlines, wind shear, pressure, and precipitable moisture, soundings of temperature and wind velocity in the upper air, a few pictures, and flight cross-sections showing prevailing cloud distributions. They argued that convective clouds are not merely decorative features of the tropical atmosphere, but are in fact the working cylinders in this critical part of the planet's heat engine. Tropical clouds also play a major role in the radiation budget and are the main water supply for 30% of Earth's surface. The study documented how cloud distribution and structure both reveal and are influenced by large- and small-scale flow patterns.

Unfortunately, the book was out of date as it came off the press. The temporal gap between the field observations of 1957 and the final publication of the results in 1964 neatly straddled the launch of the first Earth-orbiting satellites, many of them weather-related, from Sputnik (1957) to Nimbus 1 (1964).[41] A review of the 1964 cloud book by atmospheric scientist Robert Fleagle damned the effort with faint praise:

> With great effort and ingenuity Malkus and Riehl have lifted a corner of the veil that obscures the tropical atmosphere, affording a very brief and severely limited glimpse. In almost all respects (except heights of cloud tops and bases) the TIROS satellites now provide much better data in numbers that begin to match the demands of the problem. So I am left with the impression that I have been reading the daily journal written by an early explorer of the unknown new world, full of detail about the geography encountered

[39] Malkus et al., "Cloud Patterns in Hurricane Daisy," 8–9.
[40] Malkus and Riehl, *Cloud Structure and Distributions*; also an article: Malkus and Riehl, "Cloud Structure and Distributions."
[41] US National Research Council, *Earth Observations from Space*.

but in sum providing far less reliable information than one needs to know and far less than is provided on a single page of a modern atlas.[42]

Some three decades later, in a 1992 speech to University of Chicago alumni in Washington to honor Carl Rossby, Joanne located her "hot towers" work within the context of the great man's legacy, and even called him "wrong" about certain things:

> Today the legacy of Chicago's Rossby indeed extends into virtually every branch of research in atmospheric and oceanographic science. In fact, in his last few years, Rossby had turned his attention to atmospheric chemistry and did pioneering work, albeit incomplete, on such subjects as air pollutants, both man-made and natural, and was the first scientist to identify the problem of acid rain. Many of his ideas... have ushered in whole new approaches to the science and applications of meteorology, including numerical weather prediction—and such crucial concerns as the consequences of increasing amounts of greenhouse gases in the atmosphere. However, like all extra-ordinarily prolific scientists, some of Rossby's ideas were wrong, and some of his foresights missed the bus. Most notably, it turns out that Rossby's broad global perspectives of weather systems and the sources of energy that drive them dramatically missed their mark in some key aspects, as discovered by the researches of his progeny and their colleagues.[43]

Recall that Rossby once told Joanne that no-one but a little girl would be interested in working on cumulus clouds. His view of the central role of the Hadley cell circulation turned out to be an oversimplified concept, which proved to be inadequate, in itself, of maintaining a global heat balance. A closer look at what was actually occurring in tropical latitudes revealed that there was no large-scale mechanism for transferring the heated air from near the surface to upper levels where it was exported poleward. Joanne was closing in for the punch line: "The missing link—yes, you guessed it—is the cumulus cloud, or rather aggregations of enormous cumulonimbus clouds into clusters. These clusters form the Intertropical Convergence Zone; the tall clouds that carry out the upward transport part of the Hadley cell are now known as "hot towers."[44]

[42] Fleagle, Review of *The Tropical Atmosphere*.

[43] Simpson Papers, 4.10. Speech: "The Legacy of Chicago's Carl-Gustaf Rossby: From Hemispheric Weather Prediction to Global Warming" (with husband Bob Simpson at University of Chicago Washington Alumni Group luncheon), May 20, 1992. On Rossby's interdisciplinary scientific interests, see Fleming, *Inventing Atmospheric Science*, pp. 123–5.

[44] Ibid. The Intertropical Convergence Zone (or ITCZ) is a globe-circling planetary-scale area of convergence marking the rising branch of the Hadley cell.

5

UCLA

You said the full professorship would be the ultimate jewel in my crown, and that I would everlastingly regret it if I turned it down.

JOANNE SIMPSON

Decisions, Decisions

In 1959, Joanne received an offer she could not easily refuse—a full professorship at UCLA. She found the offer "staggering." Being appointed as the first female professor of meteorology in one of the premier programs in the country would be "the first really respectable position in her life."[1] Willem received a parallel offer. Joanne and Willem were earning about $8,000 each at Woods Hole, but UCLA was offering $14,000—a difference that was almost exactly balanced by the increased cost of living in the Los Angeles area.[2] Nepotism rules, which had cost Joanne her position as an instructor at Chicago in 1945, were being rescinded, at least at major universities. About the same time, Nobel Laureate Maria Goeppert-Mayer and her husband Joseph Mayer had broken through, and became full professors at the University of California, San Diego. Although she was driven to excel, Joanne vacillated about the offer for personal reasons. She was not eager at the prospect of leaving their new, larger home in Falmouth, separating their son Steven from his middle-school friends, and moving far away from their son David, who was attending boarding school in Concord. Most of all, she did not want to move so far away from Claude. Finally, after a year of indecision, the convergence of a number of factors persuaded her that it was too good an opportunity to pass up.

At the time, employees of Woods Hole were quite disgruntled after explosives specialist Paul Fye, from the Naval Ordnance Laboratory,

---

[1] Simpson Papers, 6.7. Teaching: Simpson's narrative re: teaching at UCLA, 1993.
[2] Ibid.

First Woman: Joanne Simpson and the Tropical Atmosphere. James Rodger Fleming,
Oxford University Press (2020). © James Rodger Fleming.
DOI: 10.1093/oso/9780198862734.001.0001

assumed the directorship and proceeded to impose burdensome new levels of bureaucracy on the institution and micromanage everything, from travel and small purchases to hires and promotions. Joanne called him "a failed oceanographer" and "a bureaucrat's bureaucrat." In her opinion, "when he took over, the place really deteriorated. Before him we never had to have travel orders. The navy would write some piece of paper and you'd go. And he decided we ought to have travel orders, and we ought to have permission to travel, and we ought to have permission to do this, and he put in a dictatorial deputy head of Woods Hole." This resulted in a "palace revolt," led by Henry Stommel. Fye was vengeful and made things hard for those who rebelled, after which many of the best scientists chose to leave the institution. After their collective departures, Fye closed down the aircraft research program and virtually all of marine meteorology.[3] It was a good time to move on. However, the fundamental factor weighing on Joanne's decision involved her deteriorating marriage and Willem's abuse: "I had been having a lot of difficulties getting along with my husband and he was getting quite violent. He sort of made an ultimatum, 'I want to go there and I will give you another chance to make the marriage work—you've got to go there.' So I was sort of intimidated into accepting the position, although I think I probably would have done it just because of having been a woman all my life and having been a nobody, and all of a sudden I was offered a position to be a somebody."[4]

The year 1960 also marked a turning point in Joanne's relationship with Claude. In her journal entry of January 3, she reminisced about their years together and their shared experiences, both every day— "a sentence at coffee or on the telephone, a glance, many evenings in the same spot melting together summer and winter"—and exceptional— "a typhoon in Guam, Fisherman's Wharf and the shimmering Golden Gate sunset, drifters' reef on Wake Island." Their love had grown: "I was young when I began to write here, not yet thirty, only struggling to

[3] Simpson, "Oral History," by Harper; Simpson Papers, 2.10. Clippings: "Beginning of research career in atmospheric sciences—Joanne Malkus Simpson, 1953–1964, early recognition, Imperial College, London-Guggenheim Fellowship-clippings"; Simpson Papers, 6.7. Teaching: Simpson's narrative re: teaching at UCLA, 1993; Oreskes, "Stommel"; "Brief history of meteorology at the Woods Hole Oceanographic Institution, Sept. 1957," Columbus Iselin Papers, MC-16, Box 3, Folder 9, Woods Hole Oceanographic Institution, Data Library and Archives.
[4] Simpson Papers, 6.7. Teaching: Simpson's narrative re: teaching at UCLA, 1993; Simpson, "Oral History," by Harper.

become a woman. Now I am surely entering middle age, approaching thirty-seven, and, oh my friend, my love, my brother, my son, my father, and my husband, how very much more I love you...For you and my shared world with you are my life, and to me the record of my life is the record of my love."[5] Just four months earlier, on a Labor Day outing to the Lowell Reservation on Cape Cod, Claude had said he wished to marry her. They had shared Christmas, and just three days before they had spent New Year's Eve and morning together.

> My place in your heart is no longer felt by me as conditional, to be risked if I misbehave, to be lost by absence, trauma, or trouble. I am Jo to you and there never will be another, and I am still your Jo, whether in tears or happiness, in sleep or in anger, in your arms, in Chicago, or in California. We are both often moody, both at times seized with depressions, hurt hearts, fears and death wishes, pains and hostilities inflicted each upon the other. But our acceptance of each other's lives, through them, accepts these, even loves them as a real part of a real loved one—we walk through darkness, but not alone. You are you, you are unique, you are my love, with as many facets as the oceans, as the seasons, and I choose to hold your hand through the fogs and bleak winters, even if, as usual, spring were not far behind.
>
> You are passionate, curious, and sparkling. You are undrawn, difficult, and moody—you rave on about the Catholic Church until I could slap you—you stimulate, climb to heights of love and glory, and descend to giggles and incessant chatter. You clam up and say almost nothing. You are amazingly gifted with machines and gadgets. You bring new facets to poetry and history—you are a drunken old Scandinavian, almost 50, with complex emotions, startling immaturities intermingled with gentle wisdom and loyalty, an ordinary man, but the most marvelous of man's and nature's mysteries.[6]

Joanne was facing a crucial decision—whether to stay with Claude or go with Willem. Of two minds, but ever the gentleman, Claude said he respected her love for science and her yearning for success, achievement, and recognition. He told her a full professorship was too good to pass up, and she would regret it if she turned it down. Joanne was ambivalent. She was afraid of Willem, of another broken marriage, yet ambitious to be the first female professor of meteorology. Ultimately,

[5] Simpson Papers, 1.7. Personal Diary III: 340–2, Jan. 3, 1960.
[6] Ibid., 343–6, Jan. 3, 1960.

she went, thinking she could turn back if things did not work out. But she never did. She pressed on, falling back on work as a haven from her personal problems.

Moving West as Life Goes South

Joanne and Willem arrived in Los Angeles in early 1961. UCLA's meteorology department was a national leader. It had been founded in 1940 by Bergen School legend Jacob Bjerknes. She liked her colleagues, who included Jörgen Holmboe, Morris Neiburger, and old friend Yale Mintz, the department chairman who had come from Chicago. She did not like the department's policy of hiring only one person in each area, which meant each professor was pretty much on their own, with limited opportunities for productive technical discussions with one another. A welcome exception was tropical meteorologist T. N. Krishnamurti, who came to UCLA from Chicago while Joanne was still there.[7] Joanne missed the tight-knit community she had left behind in Woods Hole. Skyrocketing real-estate prices in Los Angles had resulted in faculty residences being scattered far and wide. As a result, social interaction among department members and their families was relatively rare. Joanne and Willem's commute was about 25 miles each way, most of it along the scenic Pacific Coast Highway between UCLA and Malibu, where they lived in a luxurious house on a cliff overlooking the Pacific Ocean.

At UCLA, Joanne focused more on teaching than original research. She was expected to offer three courses per year, and had volunteered to teach the basic two-semester sequence of atmospheric dynamics and thermodynamics and a new course on physical oceanography and sea-air interactions. She was an excellent teacher who wrote out each of her lectures in detail. Since there were no adequate modern textbooks in these fields, she had to invent or reinvent, pretty much from scratch, the syllabus, readings, and laboratory exercises for each course. Joanne had learned early in her career—during the war, and especially at Illinois Tech—that the best way to learn something was to teach it. She mentored her own group of graduate students and postdocs, including William Woodley, Joe Golden, John Brown, Roger Williams, and Everett Nickerson, and kept in touch with them over their subsequent careers.[8]

[7] Simpson, "Interview," by LeMone; Simpson, "Oral History," by Harper.
[8] Simpson Papers, 6.7. Teaching: Simpson's narrative re: teaching at UCLA, 1993.

At the time, atmospheric science was growing, almost explosively, and Joanne needed to catch up on recent developments. The International Geophysical Year of 1957–58 and the Sputnik satellites triggered a chain of events that resulted in the appointment of a presidential science adviser and a new era of greatly expanded federal funding for science and technology, including meteorology. As new departments of atmospheric science proliferated, the National Science Foundation established the University Corporation for Atmospheric Research in 1959 and the National Center for Atmospheric Research in 1960. The first weather satellite, TIROS, launched in 1960, opened a new era of observations from space.[9] The availability of digital computers fostered a move away from linear theories with analytical solutions and a turn to numerical models that could, for the first time, treat the non-linearity of atmospheric processes. Joanne soon came to realize that processes she had been taught to dismiss as negligible in large-scale flows could become very important when considering small-scale phenomena such as clouds or waterspouts. Her attempt to model positive feedback between the release of latent heat of condensation and cumulus cloud updrafts was a perfect example of a basically non-linear problem, intractable to linearization.

A few months after arriving in Los Angeles, Joanne discovered she was pregnant. At age 38, after years of being told by her doctors that she was infertile, she had given up worrying about birth control. Joanne was four or five months along when she and Willem returned to Woods Hole for a visit in the summer of 1961. Claude reaffirmed his love for Joanne, but was indecisive, since he had recently become romantically involved with Margaret Chaffee, a former research associate who had worked with both of them on the data from the Pacific Cloud Hunt. Joanne's reunion with Claude enraged Willem, and he became obnoxious and abusive. He said he would keep Steven, and Joanne and Claude could have "that baby." The stress on Joanne was enormous. She loved Claude, but lacked the courage to break away from Willem.[10]

Like most employers at the time, UCLA had no maternity leave policy. In the fall semester of 1961, in her third trimester, Joanne taught Meteo 131, Atmospheric Dynamics and Thermodynamics (Figure 5.1). On November 17, three days before her due date, she gave her prepared

[9] Fleming, *Inventing Atmospheric Science*, pp. 176ff; US National Research Council, *Earth Observations from Space*.

[10] Simpson Papers, 1.7. Personal Diary III: 353, Jan. 3, 1960.

Joanne and
W. Malkus

UCLA

1961

Figure 5.1 Joanne (expecting) and Willem discussing weather maps at UCLA in 1961. Simpson Papers, 4.6. Scrapbook: "Farewell to Experimental Meteorology Laboratory of NOAA Job," 1974.

notes to Willem and asked him to cover the next two weeks of the course. Their daughter, Karen Elizabeth, was born on November 20, 1961. A note on the syllabus reads, "actually I was back by Dec. 4."[11] Unlike her mother Virginia, Joanne had always wanted a baby girl and was absolutely delighted. She recalled years later, "I certainly wouldn't ever regret having my daughter."[12]

A new baby could not save their marriage, however. In 1962, a young Swedish *au pair*, Ulla Aronsson, moved in to take care of Karen and manage the house. Willem soon became romantically involved with her and resumed his abusive behavior toward Joanne. Blaming herself, Joanne tried to justify Willem's actions. She admitted that she too had many faults in the marriage, among them seeking love and approval elsewhere. Her psychiatrist gave her vague and unhelpful advice: "Don't be ashamed. That's the way most battered wives react. They think it's their fault and are being told it was their fault and if they behaved properly and were a decent wife, why that wouldn't happen

[11] Simpson Papers, 4.12. Teaching: "Meteorology: Dynamics and Thermodynamics of the Atmosphere" (part I & II lectures, UCLA), 1961–1962.
[12] Simpson, "Oral History," by Harper.

and so on."[13] The situation worsened, and Joanne and Willem separated in 1963 and divorced the following year.

The breakup was hard on the children. David, age 19, was majoring in science at Yale at the time. Joanne had raised him as a single parent until age three. Then, Willem adopted him and formally changed his name to Malkus. He attended kindergarten and public schools in Chicago, Woods Hole, and Falmouth. In 1957 his parents sent him off to Middlesex School in Concord, where he graduated in 1963. Steven, age 14, suffered the most from the divorce. He was in his first year in boarding school back east at Middlesex. He had a rough start in life as an RH-baby, requiring many blood transfusions. Yet, within a month of his birth, his self-described "workaholic overachieving" mother had returned to her office at Woods Hole, where a babysitter would bring him twice a day to be breastfed. Joanne fought to overcome periods of guilt she felt regarding his situation, but expressed her feelings, not as a loss of opportunity to love, but rather more rationally, as a "waste of his immense talent."[14]

Willem married Ulla in 1964. By that time, Joanne had lost confidence in her abilities as a mother. To avoid complicating the situation further, she relinquished three-quarter time custody of Karen to Willem and Ulla, since she saw Ulla as being a stable, loving, and outgoing stepmother and housewife.[15] As a further insult, Joanne's mother kept a picture in her album of the wedding of Willem and Ulla in Sweden, accompanied by a note saying that she had retained great affection for Willem and took his side in the breaking up of his marriage. Joanne was left destitute. Willem gained possession of their big house in Falmouth, all of the family albums, and most of Joanne's possessions: "I lost essentially any record (except that of work) of 15 years of my life (1948–1963). This was an important time because of my sons, all of their childhood and some of their adolescence, and the first two years of my daughter's Karen's early childhood."[16]

[13] Ibid.

[14] Simpson Papers, 1.16. Family history: Simpson's narrative, January 1998, re: marriages, children, and other family from 1940s to 1970s, and captions for family photographs.

[15] J. Simpson, "Meteorologist."

[16] Simpson Papers, 1.16. Family history: Simpson's narrative, January 1998, re: marriages, children, and other family from 1940s to 1970s, and captions for family photographs; Simpson Papers, 2.8. Clipping scrapbook, 1947, 1961–1973 (compiled in 1999).

Here are Joanne's personal reflections on the turmoil surrounding her decision to go to UCLA, written in 1984 after receiving the news of Claude's passing:

Last night I cried for the first time since you died. I couldn't stop crying—and I don't often cry now, it tears [me] up too much...How did it all come out the way it did? In novels, lovers either marry or leave each other in bitterness and estrangement. I have to try to trace out the agonies of the early sixties, the divorce, where you alone stood by me and up for me, the tangled agony with Margaret, my marriage to Bob (did that, which worked out so well begin as a rebound, or more, a rebound from its own initial agony and tangle), to when again we found ourselves close, as close in mind, thought, and taste as ever, but platonic friends, only kissing on arrival and departure. The last time I wrote in this notebook in 1960 was near the last time I believed you really loved me, and after that, things went into a terrible tangle. I believe now, a generation later, that the only chance we had to spend more years as lovers was if I had had enough stubbornness and courage to refuse to leave Woods Hole for UCLA. Let Willem go if he wanted. If you, my dear, had lifted a finger or [uttered] a sentence to persuade me, I would not have gone but stayed with you—then there would have been no baby Karen and no Mad Margaret, and Joanne and Robert Simpson would not be celebrating their twentieth wedding anniversary on December 29, 1984. Am I sorry? Were you sorry? Did you ever wish you'd asked me to stay? I'll never know.[17]

Joanne had little time to produce new research during her four turbulent years at UCLA. Two articles on Hurricane Daisy appeared in *Tellus* in 1961. One she published with Riehl; ironically, the other was coauthored by Claude Ronne and Margaret Chaffee. Her new course on physical oceanography resulted in a long review chapter on large-scale interactions in the ocean environment.[18] Together with UCLA graduate student Roger Williams, she improved the computer model of convection supporting her hot towers thesis.[19] She again crossed the Pacific Ocean to lecture on tropical convection at a symposium sponsored by the World Meteorological Organization in Rotorua, New Zealand, in 1963.[20] Her long-delayed book with Herb Riehl, *Cloud Structure and Distributions over the Tropical Pacific Ocean*, finally appeared in 1964.

[17] Simpson Papers, 1.7. Personal Diary III: 348–54, Nov. 1984.
[18] Malkus, "Large Scale Interactions."
[19] Malkus and Williams, "On the Interaction Between Severe Storms."
[20] Malkus, "Tropical Convection."

Joanne's professional accomplishments earned her the Clarence Leroy Meisinger Award of the AMS, "for outstanding experimental investigation of cumulus clouds by means of aircraft measurements and studies of atmospheric circulation in the tropics." During this time of personal misery and turmoil, she received one of nine "Woman of the Year" awards by the *Los Angeles Times*. The article smoothed out all of her rough edges, as tends to happen in basic biographical sketches designed to appreciate a person's resume-worthy accomplishments. The article did not mention her first marriage to Victor Starr or her failing marriage to Willem Malkus. Ironically, the article featured a photograph of her at home with Willem, Karen, and their collie MacDuff in a scene of apparent domestic tranquility. The article described her hobbies as "loving to cook and to sew her own clothes," and only later, "casual flights for cloud observation." This sort of elision and personal falsification was not uncommon at the time due to the ever-present respectability politics that served as a determinant of who should be societally validated. If her home appeared to be in shambles, or if divorces made her appear to be operating outside of the nuclear family model, people might write off her life as a failed attempt to balance home and work.[21]

"Hurricane" Bob

Seeking comfort, sanity, stability, new challenges—and yes, love— Joanne turned her attention away from Los Angeles and to hurricane forecaster Bob Simpson, a long-time professional associate and kindred spirit. They first met in 1943 at a seminar in Chicago, then worked together for several years under the auspices of the NHRP. In 1959 the weather bureau sent Bob to the University of Chicago to complete coursework for his doctorate. He enrolled in Herb Riehl's seminar on tropical meteorology, which Joanne was auditing. Claude visited her in Chicago. She and Claude and Bob spent several "wonderful weeks" together in that late winter and in early spring. Bob was going through a divorce, and, Joanne recalled, they became very close. After Claude left, "I was there all by myself, and he was there all by himself . . . He was working pretty hard, but we got to know each other more then."[22]

[21] Bengelsdorf, "Woman of the Year."
[22] Simpson, "Oral History," by Harper; Simpson Papers, 1.7. Personal Diary III: 348–54, Nov. 1984.

Joanne, Bob, and Herb Riehl discussed issues of scale and the possibility of intervening in hurricanes through cloud seeding. Did their movement and development depend solely on conditions in the surrounding environment, or did they respond to smaller-scale changes in the cumulus clouds? Joanne focused on the clouds, Bob emphasized the mature hurricane itself, while Riehl favored the large-scale environment. Bob hypothesized that seeding the eye wall clouds of hurricanes with silver iodide might enhance the vertical motion and enlarge the eye, spreading the energy of the storm out over a larger area, and potentially reducing the intensity of its winds. In 1958 the NHRP had attempted to seed Hurricane Daisy, but the effort was unauthorized and the results were inconclusive. To test his theory, Bob needed a way to experiment on hurricanes and measure the results.[23]

At Chicago, Roscoe Braham's lab had a contract to develop airborne measuring instrumentation and cloud-seeding equipment for hurricane studies. The saturated conditions in hurricanes, however, made the existing silver iodide–acetone burners unusable, so the project had hit a technical snag.[24] "All of a sudden a guy turned up from Naval Ordnance out in China Lake who had invented silver iodide flares. This immediately hit Bob as a great idea. He had a theory of how to seed hurricanes to reduce their wind force."[25] The year was 1960 and the "guy" was Pierre St. Amand. The flares, called Alecto after one of the Greek Furies, offered an exciting new way of seeding a cloud or a storm by dropping pyrotechnics into it from above.

In 1961 the NHRP dropped some of these flares into Hurricane Esther, which displayed some apparent weakening after seeding. This encouraged the weather bureau and the navy to develop an official and aggressive hurricane modification project. Such is the origin story of Stormfury, a project sponsored by the National Science Foundation and conducted by the weather bureau and the navy.[26] The navy offered airplanes, airborne radar, flight crews, and the expertise needed to coordinate the flight paths of up to thirteen aircraft. Cooperation between the weather bureau and the navy meant the project had two

[23] R. H. Simpson, *Hurricane Pioneer*, pp. 96–8.

[24] Braham. "Interview," by Cole; Braham and Neil, "Modification of Hurricanes."

[25] Simpson, "Interview," by LeMone.

[26] Willoughby et al., "Project STORMFURY: A Scientific Chronicle, 1962–1983"; US Department of Commerce, National Oceanic and Atmospheric Administration, *Project Stormfury*.

sponsors; but it also had two agendas. The weather bureau aimed for better understanding of hurricanes and a reduction of their strength by seeding them; the navy wanted to control the air–sea environment, calm heavy seas, redirect violent storms against potential enemies, and even modify the climate, as needed, for the sake of military operations.[27] St. Amand, in particular, sought to intensify and control hurricanes, certainly for tactical advantage, but also perhaps as weapons of war. Bob Simpson recalled, without specifying the details, that St. Amand did not share his scientific values and "succeeded in throwing monkey wrenches into the works."[28] In 2007, Joanne Simpson, then a retired NASA employee, recalled in an on-camera interview, "I thought it was terrible—I mean all my life I've tried to work for the betterment of the planet and the people in a small way— and to use what I have done as some kind of a military thing. I obviously am very concerned and not happy about it."[29]

Joanne styled herself an antagonist, not a protagonist on the project. She was pessimistic about the chances of successfully controlling hurricanes, largely because of their huge natural variability, but was very hopeful about what could be learned through experimenting on them. She thought that the idea of attempting to make major changes in a hurricane's eye wall was highly speculative, given the huge amounts of energy that the storm released. Nevertheless, she wanted to see how clouds and hurricanes responded to perturbations by massive seeding and to compare their subsequent behaviors to her models.

Experimenting on hurricanes was exciting, somewhat dangerous, and politically charged. Bob's deputy, R. Cecil Gentry, referred to them as "the largest and wildest game in the atmospheric preserve... with urgent reasons for 'hunting' and taming them."[30] In October 1962, just as Project Stormfury was getting under way, the Cuban missile crisis brought the world to the brink of nuclear war. The following year, Fidel Castro accused the United States of weaponizing hurricane Flora. The storm hit Guantánamo Bay as a Category 4 storm and made a 270-degree turn, lingering over Cuba for four full days, with intense driving rains that caused catastrophic flooding, thousands of deaths, and extensive crop damage. The weather bureau insisted that Flora had not been seeded, although suspicions lingered that it had. Their official

[27] Fleming, *Fixing the Sky*, pp. 178–9; US Navy, "Technical Area Plan," 1.
[28] R. H. Simpson, "Interview," by Zipser.
[29] "Science of Superstorms."
[30] Gentry, "Hurricane Modification," 497.

guidelines were not to intervene in any hurricane that might hit land within 36 hours. This rule so severely restricted the experimental area in which seeding would be allowed that, according to Bob, "little hope remained to demonstrate statistically that hurricanes could be usefully seeded."[31] It was a media circus, with confrontational reporters incessantly demanding stories. They suspected something covert was going on, especially when Joanne clammed up and refused to grant them interviews.

Joanne had volunteered to go on Stormfury flights to pursue a scientific objective of her own—experimentation on cumulus clouds to compare their behavior with that predicted by her computer model. Since, by then, there was no restriction on women going on flights, she found it very exciting and easy to participate in these missions and work with the data.[32] Her 1963 "Stormfury cumulus experiments" succeeded beyond her wildest dreams. Nearly all the seeded clouds grew lustily compared to initially similar ones that were designated as controls. Most of the ten or so clouds they seeded grew spectacularly compared with the surrounding clouds. Joanne recalled, "When that first seeded cloud exploded, I was never more excited in my life. All the scientists and crew of the several aircraft were cheering, by radio and intercom."[33] Her simple cloud model also functioned well, predicting the behavior of the seeded and unseeded clouds.

Later that summer, Bob and Joanne experimented on hurricane Beulah, bombing it with seeding flares with results in line with the Stormfury hypothesis: "By 'seeding' a cluster of clouds near the center of a hurricane with silver iodide crystals it may be possible to trigger a self-sustaining chain of events leading to a reduction in the storm's wind speed."[34] The foremost link in the chain was that the storm core had to have supercooled liquid water, which could be converted into ice. They presented a rationale for interference, a quest to find the "Achilles' heel" that might hobble a monster storm, an argument for more experimentation, with emphasis on the conditional language, "may be possible." When seeded, Beulah's existing eye wall disappeared from the radar screen, a new one formed nearly 20 km farther from the storm center, and the maximum wind speed decreased by 15%. Perhaps

[31] R. H. Simpson, "Interview," by Zipser.

[32] Simpson Papers, 3.12. Narrative (by Simpson): "The Miami Years, 1967–1974."

[33] Ibid.; J. Simpson et al., "Stormfury Cumulus Experiments."

[34] R. H. Simpson and J. S. Malkus, "Experiments in Hurricane Modification."

they were learning how to tame a hurricane, perhaps the seeding had caused smaller raindrops or ice crystals to form, both invisible on radar, or perhaps it was an epiphenomenon caused by instrument limitations. Frustration within the program grew as the Stormfury scientists began to realize their hurricane-seeding hypotheses were flawed. First of all, meteorologists assumed (later found to be in error) that hurricanes contain very little supercooled water necessary for effective silver iodide seeding. Joanne wanted to practice experimental meteorology by adding small chemical perturbations to a system and then measure the dynamical response. The effects of seeding were so small and the variability of the environment so large, however, that interventions were impossible to measure.

Bob received his doctorate from Chicago in 1962 and returned to the weather bureau to direct Stormfury, a project that fascinated Joanne and brought her back into his orbit. They published several articles on their hurricane modification hypothesis and cloud seeding trials.[35] One article appeared in *Science* with a photographic collage on its cover depicting trade cumuli, exploding clouds, and silver iodide flares.[36] Joanne called this the beginning of "the Malkus/Simpson collaboration and increasingly close friendship." In 1948 Irving Langmuir had written about "chain reactions" in cumulus clouds, Joanne and Bob were now "exploding" them using silver iodide bombs.[37] They claimed they had taken the first steps in making meteorology into an experimental (and possibly operational) science by conducting relatively controlled and theoretically modeled experiments on cumulus clouds. As it was in Langmuir's case, this claim was controversial, and generated an immense storm of opposition: "We were attacked from many sources, some of them pretty nasty. I could not believe what was happening. I had written some speculative papers before, some with Riehl and some on sea–air interaction, and no one complained, except for an occasional comment from someone who said he did not agree with part of section so-and-so."[38] Even her 1963 "Woman of the Year" award from the *Los Angeles Times* was a source of controversy, since her colleagues did

[35] R. H. Simpson and J. S. Malkus, "Experiment in Hurricane Modification; R. H. Simpson and J. S. Malkus, "Experiments in Hurricane Modification."

[36] J. S. Malkus and R. H. Simpson, "Modification Experiments on Tropical Cumulus Clouds."

[37] Langmuir, "Production of Rain by a Chain Reaction."

[38] Simpson Papers, 3.12. Narrative (by Simpson): "The Miami Years, 1967–1974."

not appreciate her explaining the Stormfury project in a newspaper article before any peer-reviewed reports appeared. Joanne naively thought her colleagues in atmospheric science would be pleased to see the good performance of the model, and she struggled for metaphors to describe the backlash: "The fat was in the fire. Within the next few months both my back and front were filled with daggers, mud, dead cats, and rotten eggs. My polite colleagues would say 'Excuse me, I don't believe you'."[39] Joanne found it hard to get any work done in a politically charged atmosphere where there was so much hassle and so much unpleasantness. She spent more time dealing with politics and the press, putting out the fires, than trying to learn about hurricanes.[40]

To cool things down and not disgrace the UCLA meteorology department and themselves, Bob and Joanne assembled an informal advisory group consisting of the UCLA department chairman, Yale Mintz, and several experienced weather bureau scientists, the key individual being statistician Glenn Brier, who had worked with scientifically-based weather modification experiments before. The group advised them that they had to do two things to gain credibility. The first was to submit for publication reports on the hurricane-seeding hypothesis, its basis and progress so far. It is a negative in science to have anything in a newspaper before having it sent to and, better yet, approved by a peer-reviewed journal. Joanne and Bob published papers on the cloud model in the *Journal of Applied Meteorology*, *Oceanus*, and elsewhere.[41] They repeated the cumulus part of Stormfury as a double blind experiment with decisions to seed or not to seed made by an outside statistician.[42]

In 1960, as Joanne was agonizing over the decision to leave Woods Hole and go to UCLA, Claude had said to her that the full professorship would be "the ultimate jewel" in her crown and that she would everlastingly regret it if she turned it down. Hesitantly, Joanne went, and, following a pattern established in childhood, her work became a retreat

[39] Simpson Papers, 18.10. Simpson's narrative: "My Experience with Weather Modification," March 2006.
[40] "Science of Superstorms."
[41] J. S. Malkus and R. H. Simpson, "Note on the Potentialities"; Malkus and Simpson, "Ocean as a Laboratory."
[42] Simpson Papers, 18.10. Simpson's narrative: "My Experience with Weather Modification," March 2006.

from and recompense for all her personal problems. She wrote in her diary to Claude, "You really were of two minds whether you wanted me to go or not—and I was afraid of Willem, of another broken marriage, and ambitious to be the first female professor of meteorology and I went, thinking I would turn back after a year or less—but then it was too late. I wrote you, but you didn't answer, I don't know why."[43] By 1964 Joanne realized that the full professorship at UCLA had not turned out as planned. Instead, her experiences there were emotionally devastating, and she decided to ease her personal turmoil by taking a new direction and taking on new professional challenges. She wanted to fly through the clouds and conduct experiments on them, and felt that she could do that best by going into the weather bureau, where Bob worked.[44]

[43] Simpson Papers, 1.7. Personal Diary III: 348–354, Nov. 1984.
[44] Simpson, "Interview," by LeMone.

6

NOAA

Meteorologists want to understand how clouds work, not how
clouds can be put to work.

JOANNE SIMPSON

A Fresh Start

Joanne visited Woods Hole in the summer of 1964, seeking to put her
shattered life back together. She was still an associate meteorologist
there and was serving on a National Academy panel on weather modification. Robert M. White, the new Chief of the US Weather Bureau,
was in town for a meeting. They discussed common interests: tropical
convection, her computerized cloud model, her opinion of recent
experiments aimed at modifying hurricanes, and her role as adviser to
Project Stormfury. White invited her to consider her next steps: "Well
you know, we've gotten more and more involved in pressures to do
something about weather modification, and I need somebody to be a
sparkplug of that. How would you like to come and work for the
weather bureau?"[1] Although she held a dim view of bureaucratic
organizations, Joanne jumped at the chance to move to Washington
as a consultant. She was following her heart, "with the purpose of
enabling Bob and me to get married; otherwise I would have never left
a full professorship for the weather bureau."[2] During her tenure there,
large-scale administrative changes resulted in the weather bureau being
folded into ESSA (Environmental Science Services Administration) in 1965
and into NOAA (National Oceanic and Atmospheric Administration)
in 1970.[3] The bureaucracies became immense.

On December 29, 1964, Joanne, age 41, and Bob Simpson, age 52, melded
their lives and their careers in a partnership that lasted for the next

[1] Simpson, "Oral History," by Harper.
[2] Simpson Papers, 3.12. Narrative (by Simpson): "The Miami Years, 1967–1974."
[3] Popkin, *Environmental Science Services Administration*; White, "Making of NOAA."

First Woman: Joanne Simpson and the Tropical Atmosphere. James Rodger Fleming,
Oxford University Press (2020). © James Rodger Fleming.
DOI: 10.1093/oso/9780198862734.001.0001

forty-five years. It was the third marriage for each of them. They exchanged vows at the county courthouse in Arlington, Virginia, and then celebrated a church wedding the following week in Coral Gables, Florida. Bob hit it off immediately with Joanne's children, David and Karen, although 14-year old Steven was quite reserved. Bob had two grown daughters: Peg, 26, and Lynn, 22, who remained close. Joanne and Bob first rented a fifteenth-floor apartment in Rockville, Maryland, then moved to Annapolis within the year and acquired their first sailboat, a Bristol 27 which they christened *Sabrina*.

Before they married, Joanne and Bob developed a professional relationship centered on hurricanes and tropical meteorology. She referred to them as a "terrific research team with complementary skills. Where I was weak, he was strong, and vice versa. The two of us together were able to create more than twice as much as the sum of our individual efforts. In part we married because we looked forward to a long life of working closely together. That was the daydream."[4] In reality, the new couple immediately faced unwelcome nepotism rules imposed by the federal civil service, since the weather bureau could not employ both of them in research.

Experimental Meteorology

Robert White asked Joanne to head the Experimental Meteorology Laboratory (EML), then located in Washington. He also appointed her interim director of Project Stormfury, a position she was reluctant to assume because of the politics of weather control. Many of her colleagues were vehemently opposed to any attempts to seed hurricanes. Joanne recalled: "I was afraid it would take too many experiments to really demonstrate a seeding effect. But I had accepted a job in order to be near Bob and was thereby stuck with it."[5] Bob was moved out of his research position into a desk job as director of operations for the weather bureau, a high-level administrative position. He remained a consultant to Stormfury, as Joanne had been earlier, and they were able to take research flights together. For the next five years, however, the hurricanes failed to cooperate, with no storms deemed suitable for seeding. After intervening in only two hurricanes, Joanne and Bob still

[4] J. Simpson, "Meteorologist."
[5] Simpson Papers, 3.12. Narrative (by Simpson): "The Miami Years, 1967–1974."

had a long way to go to test their hurricane hypothesis and reach their
goals of being able to deflect them, reduce their destructive power, or
even understand them. Although they had better equipment—aircraft,
flares, and radar—they were overly optimistic in their claim that Stormfury
was providing them with the ability to perform actual experiments in a
full-scale atmospheric laboratory in order to develop and test various
modification hypotheses.[6] In reality they were flying in a turbulent,
tropical environment buffeted by some of the strongest winds on the
planet—hardly controlled laboratory conditions. Yet they remained
confident, certain even, that weather forecasting in general could be
greatly improved by their experimental methodology of applying what
they called "a known force" to an atmospheric phenomenon and measuring the results, rather than patiently waiting to observe only what
nature chooses to reveal.

Joanne's pathway into experimental meteorology stretched all the
way back to her early numerical models of cumulus convection. In
1954, while at Imperial College on her Guggenheim Fellowship, Rossby
encouraged her to abandon the laborious practice of hand calculations
using a slide rule and work with him in Stockholm using the BESK
computer. In addition to modeling heated plumes of air, she and colleague Georg Witt included subroutines representing the release of
latent heat as water vapor condensed into cloud droplets and ice
formed in the rising updrafts. Later, Joanne returned to her model to
see what it revealed about the behavior of a seeded cloud that might
suddenly glaciate. Would a seeded cloud grow faster and rain more
than its unseeded counterpart? Her model indicated that it would.
Joanne focused on the science rather than the utility of weather
modification. As she expressed it: "From years of study I had learned
that the life cycle of a cumulus cloud is a fierce struggle for existence,
with the forces of growth and destruction in near balance. Because
it is difficult to study a nearly balanced natural phenomenon, I wanted
to upset the balance."[7] Joanne wanted to use seeding field trials as
experimental tools to gain insights from her cloud model rather than
using seeding as a way to get immediate benefits, such as increasing
rain or reducing hail damage. She wanted to know how clouds work,
not how clouds could be put to work.

[6] R. H. Simpson and J. S. Malkus, "Experiments in Hurricane Modification," 37.
[7] Noble, "Joanne Simpson, Meteorologist," 41.

Joanne's cloud model predicted that in the tropics, most of the seeded clouds would grow taller compared to those that were unseeded. Experimenting on clouds would allow her to test multi-aircraft coordination in fine weather before trying it in a murky hurricane. As she had done a decade before, Joanne flew out of Puerto Rico with her old friend Claude Ronne along as the photographer. She was authorized to conduct a few cumulus experiments while the rest of the team waited for a seedable hurricane. Joanne expected the scientific community to recognize this work in bringing together cloud experimentation and modeling. According to her, the most we got from the most knowledgeable was, "So you have shown that you can make clouds grow taller, so what?" What we wished someone had said was, "Your crude model worked well for a simple variable such as cloud height, now there will be resources available to take and test the next steps toward understanding and simulating cloud motions and rainfall." The main criticisms, coming from mainstream atmospheric scientists, included reasonable requests to provide more details about the seeding experiments and to clarify the claim that seeding was a "success."[8] Joanne was ready to leave the contentious field of weather control behind, but Bob's fascination with hurricanes and their possible modification encouraged her to continue.

In 1965, she and Bob repeated the earlier cumulus seeding experiments, and was able to produce, results in line with the predictions of her model. It was Bob's hypothesis that silver-iodide seeding of the supercooled (below $-40°$ C) liquid water in a hurricane could be used to control the storm. In the late 1940s, during one of his pioneering hurricane research flights into the top of a hurricane at 40,000 feet, Bob had written: "[As we were traveling] at 250 miles per hour through [the cirrostratus ice] fog, there loomed from time to time ghost-like structures rising like huge white marble monuments... These were shafts of supercooled water, which rose vertically and passed out of sight overhead. Each time we passed through one of these shafts, the leading edge of the wing accumulated an amazing extra coating of rime ice."[9]

These "marble monuments" Joanne later christened "hot towers." Bob thought that seeding the hot towers in the storm's eye wall might

[8] J. Simpson et al., "Experimental Cumulus Dynamics"; J. Simpson et al., "Stormfury Cumulus Experiments."

[9] Simpson Papers, 18.10. Simpson's narrative: "My Experience with Weather Modification," March 2006.

cause them to glaciate, enlarge the eye wall, and weaken the circulation without decreasing storm rainfall. Even a 10% reduction in wind speed would decrease wind damage by at least 20%, which would be an immense benefit. The seeding of hurricanes Esther in 1961 and Beulah in 1963 produced promising but inconclusive results along these lines, but the observed changes were well within the range of natural variations in unseeded hurricanes. Joanne was skeptical that seeding had actually caused the changes, but was in favor of continuing the experiments. The promise of better understanding, improved forecasts, and the slim chance of learning how to reduce hurricane intensity far outweighed the costs of the experiments.

In 1965, the press hyped ESSA, the brainchild of Robert White, as an activist superagency with huge utopian expectations: "Accurate world-wide weather forecasts weeks in advance, solutions to the air and water pollution problems, better warnings of storms and other natural hazards, even earthquakes, human control of weather and climate, including sure-fire rainmaking and storm rerouting. [According to its administrator, Robert M. White], if these dreams of environmental scientists ever come true, they will be realized faster and less expensively because of the creation of the month-old Environmental Science Services Administration."[10]

That very summer, the dream of weather control became a public relations nightmare. In August, as hurricane Betsy traveled on its long, looping path, the Stormfury team was in the air, ready to intervene. The National Hurricane Center (NHC), however, identified a small trend in storm motion that might steer it toward the coast. When discussing the situation by telephone from the aircraft, all parties, including White, agreed that the experiment should be called off. The *Washington Post*, however, was not properly informed of this decision and erroneously reported that attempts to "tranquilize" Betsy by seeding it may have caused the storm's erratic track. The *New York Times* reported that weathermen had ordered the "bombardment" of hurricane Betsy in an attempt to cool it off, break it up, and destroy it.[11] Betsy made two landfalls—one in the Florida Keys and another on the coast of Louisiana, resulting in eighty-one fatalities and well over $1 billion in damages. Joanne was left to pick up the pieces and deny any weather bureau

[10] Downie, Jr. "New Agency a Step Toward Weather Control."
[11] "Tranquilizer Prepared"; "Seeding of Storm to be Tried Today."

responsibility; she also had to deny any Stormfury successes.[12] Continued harassment from skeptical colleagues and uninformed critics, combined with pressure from Robert White to accelerate the pace of the experiments and institute practical weather control, led to Joanne's decision in 1966 to step down from directing Stormfury. She thought the project was better suited for the National Hurricane Research Laboratory. Instead, she focused EML's efforts on cumulus cloud seeding and modeling experiments.[13]

Florida

Bob, who was largely constrained to administrative functions in Washington, missed his direct engagement with severe storms, and so took a pay cut to move to Florida in 1967 as the director of the NHC. Magnanimously, Robert White then transferred Joanne to Coral Gables to EML's new facilities on the campus of the University of Miami. EML was co-located in the computer center building, with the NHC and the department of atmospheric science on different floors (Figure 6.1).

Joanne and
Bob

1968

NHC

Figure 6.1 Joanne and Bob in 1968 working together in Miami. Simpson Papers, 4.8. Simpson Symposium, February 9–13, 2003.

[12] R. H. Simpson, *Hurricane Pioneer,* 103; Simpson Papers, 18.10. Simpson's narrative: "My Experience with Weather Modification," March 2006.
[13] Simpson Papers, 18.10. Simpson's narrative: "My Experience with Weather Modification," March 2006.

Joanne used this time to strengthen family ties—her adult children harbor no regrets about their upbringing. The Simpsons settled into a new house in South Miami, with a pool, screened-in patio, and plenty of room for guests. In anticipation of David's graduation from college and Steven's graduation from high school, they acquired *Sabrina II*, a Morgan 34 sailboat that could sleep five.[14] Their dinghies were always named *Karen*. Joanne recalled: "We had the clouds and thunderstorms we loved right in our back yard. In our front yard were Biscayne Bay, the Florida Keys, and our own sailboat to explore them."[15] In 1969 Bob and Joanne built a new larger house in Coral Gables on a protected waterway emptying into Biscayne Bay. They also bought *Sabrina III*, a Gulfstar 43 sailboat that could sleep ten and came equipped with all the amenities.

Joanne's divorce agreement allowed Karen to spend summers with her mother, and the family engaged in a lot of outdoor activities. She and her brother Steven loved to sail and did so often with their mother and Bob.[16] Steven recalled these pleasant times, but also the way Joanne took charge when their boat was in danger. "We were sailing on the Cape in a fog and entered a cove. My mother somehow knew they had gotten into a dangerous, rocky place and shouted, 'Everybody get down on the deck flat!' while she made sure everything was safe."[17] This ability to take charge impressed Steven, who characterized his mother, years after her demise, as "a remarkable person; determined, honest, and clear, who worked really hard and went through a lot of crap... Her curiosity, expressed as a burning fire within her, kept her going at all times." Steven recalled her motherly qualities, her cooking, and her attentiveness to the children. "She showed up at my graduation from Harvard and my MFA dance performance at Arizona State, and did not just say 'Oh that's great'—she actually showed up and was attentive." Karen remembered that her mother came to help for a week after she had given birth to her son Michael. "She cooked fancy desserts and made sandwiches and was great taking care of the baby... She always said she wished she could have spent more time with her children, but

---

[14] Simpson Papers, 1.16. Family history: Simpson's narrative, January 1998, re: marriages, children, and other family from 1940s to 1970s, and captions for family photographs.

[15] Simpson Papers, 3.12. Narrative (by Simpson): "The Miami Years, 1967–1974."

[16] Simpson Papers, 1.16. Family history: Simpson's narrative, January 1998, re: marriages, children, and other family from 1940s to 1970s, and captions for family photographs.

[17] Malkus, Steven, personal communication with Talia Gebhard, Nov. 27, 2018.

I never felt that that was an issue. I knew she was doing what she loved, and she was passionate about it, and obviously good at what she did. It was a great opportunity to share some personal time with her and admire her as a mom. I hope in that experience I shared with her how much of a wonderful mother she was because she sometimes doubted that."[18]

Joanne said she was happy in Miami, but later wondered why. Most likely it was due to a combination of factors: working closely with Bob, enjoyable reunions with her children, the success of her cloud model, the strong team of researchers she was nurturing at EML, and the fact that she had distanced herself from the Stormfury hurricane program. Nevertheless, she was deeply depressed and was constantly fighting the feeling of not wanting to get up in the morning. Her visits to a psychiatrist did not help at all until he prescribed the anti-depressant medication Tofranil (imipramine hydrochloride), which Joanne claimed "made the difference between survival and non-survival."[19] Possible side effects of the stimulant included mental or phys-ical hyperactivity, which would have been close to normal for Joanne. She was also experiencing the most debilitating migraine headaches of her life, and spent a week in Massachusetts with the doctor who had treated her mother. She attributed the headaches to inheritance, but blamed the severe depression on post-divorce shock and grief. At the time, she had to keep her treatments confidential, writing, "The stigma that attaches to even a few visits to a psychiatrist, added to being a woman and being divorced, were more than I thought my career opportunities could stand.[20] Karen recalled the amazing contrast between her mother's manic and depressed states: "She was this passionate, involved, excited person ... and then there were times when she was really ill, sick, and in the dark place. I say dark because physically she was in a dark place; the room was dark, the curtains were closed, she would often have an oxygen mask on or be under a towel or something to keep her comfortable ... She would be in miserable pain."[21]

In 1969, Joanne traveled to the Soviet Union as part of a US delega-tion on weather modification. The depressing conditions there came as

[18] Malkus-Benjamin, "Video Interview," by Lipshultz.
[19] Simpson Papers, 3.12. Narrative (by Simpson): "The Miami Years, 1967–1974."
[20] Simpson Papers, 1.14. Family history: Simpson's narrative re: difficult childhood, lifelong depression, detailed photograph captions with commentary, January 1996.
[21] Malkus-Benjamin, "Video Interview," by Lipshultz.

a shock to her. She had kept an open mind about "red communism" as reported by the press during the McCarthy era, but soon discovered that there was no communism at all in Russia. Rather, she discovered that their system was a huge and oppressive bureaucracy—a crude and crass dictatorship where the bosses ruled and stamped out any disagreements. The average person was fearful of expressing any criticism of the system. Consumer goods were scarce, the hotel food inedible, and the dwellings tacky. Her Russian colleagues, on the other hand, were wonderful, warm, and friendly people, who royally entertained their group. Soviet women in meteorology were scarce, and those she met had been "liberated" to serve as drudgework assistants to men in the office and to their spouses at home. Their children were not accommodated in state-run daycare centers, but nearly all were at home being cared for by grandparents. When she flew back to Copenhagen, she found herself in tears to be back in the world "free of Big Brother."[22]

Ironically, although Joanne had chafed at the bureaucracy at Woods Hole, she still had to deal with it in her professional life. In 1970 the weather service became part of NOAA, a huge new environmental agency established by an Act of Congress to incorporate oceanic and atmospheric research and service. NOAA was the largest organization, with the largest budget, within the Department of Commerce. Robert White became its first administrator and was elevated to Undersecretary of Commerce with his own chauffeured limousine.

Joanne's job at EML as a government scientist in the lower echelons of management involved issuing orders, complying with regulations, and political maneuvering in the contentious area of weather manipulation. She wondered about the basic incompatibility between the training of a research scientist and the qualities of a "good soldier" to be rewarded within a civil service organization. She had never been promoted at NOAA, but received multiple honors and recognition during this period. She received the Department of Commerce silver medal in 1967 and its gold medal in 1972, Fellowship in the AMS in 1968, a NOAA distinguished authorship award in 1969, and in 1972, listing in *Who's Who of American Women*, and appointment to a summer term at Oregon State University as a distinguished visiting professor. From 1967 to 1974 she served as an adjunct professor of atmospheric sciences at the University of Miami.

[22] Simpson Papers, 3.12. Narrative (by Simpson): "The Miami Years, 1967–1974."

What kept her going was her fierce loyalty to the careers of the fifteen or so young scholars in her laboratory, from high school to postdoctoral level, who flew with her and produced dozens of research papers, public talks, and popular articles. She cultivated what she called a "quasiacademic atmosphere" in which her protégés could develop their talents freely and fully. She fought "like an alley cat" to secure their advancement and honors, and figuratively "spilled her blood" to shield them from the nightmare of the bureaucratic and political quagmire surrounding them. She granted many interviews, often with female reporters who inevitably asked: "How does a woman fare being the boss of so many men? Don't some or all of them resent it?" Joanne minimized this problem by selecting very young men to work for her and by delegating authority and allowing each person as much autonomy as possible.[23]

Florida Area Cumulus Experiment

In 1967, recognizing the long hard road ahead for adequate understanding of hurricane modification, Joanne returned to experimenting on clouds and comparing the results to her model. Her personal favorites were cumulus clouds, where her professional career had begun. Major changes were in the air, however, in weather modification research. Her new employee, William (Bill) Woodley, claimed, in his 1969 doctoral thesis at Florida State University and in subsequent publications, that silver iodide pyrotechnic seeding of supercooled Florida cumuli induced cloud growth and may have resulted in increased rainfall. However, the EML team thought that routine use of this technique would be premature.[24] In an area south of Lake Okeechobee, Joanne and her team improved an existing network of rain gauges and used it to calibrate the University of Miami's surface radar measurements. This enabled Woodley to obtain the rain rate and total rain amount for each experimental cloud, both seeded and control, randomly chosen. The seeded cloud population appeared to produce nearly double the rain amount of the control population. Joanne and Bill floated the notion of "cumulus mergers." They observed that one cloud over Florida sometimes "merged" with another nearby to produce a larger system that

[23] J. Simpson, "Meteorologist."
[24] Woodley, "Effect of Airborne Silver Iodide"; Woodley, "Precipitation Results," 255.

lasted much longer than an isolated cloud and might produce more rain. Could they cause clouds to merge on command? Could they use cloud seeding to increase the rain over an extended area in south Florida?[25]

Joanne wanted to understand clouds from the inside out—how their microphysical properties, temperature profiles, and liquid water content interacted with the larger dynamical environment and any rain they may produce.[26] Every morning, in preparation for field trials later that day, Joanne ran her cloud model on the university's mainframe computer. The model predicted the "seedability" of that day's clouds. Seedability was the difference between the predicted top of a model seeded cloud and an unseeded cloud. A large value, of several kilometers or more, indicated that conditions were ripe for afternoon seeding.[27]

Joanne loved the flights and working on the data, but it was clear that the evaluation of weather modification experiments required strict statistical controls. It was also clear that a purely random selection of clouds was not appropriate. A big obstacle was the limited access to the research aircraft. They had only five to ten cases per year for area seeding. However, in a few years the results were looking quite positive. The rainfall appeared to be increased by the seeding, but the results were not statistically significant.

To help with statistical decision-making, Joanne and Bob applied for, and were granted, a one-semester sabbatical at the Thayer School of Engineering at Dartmouth College to study decision analysis and to learn to use or at least be aware of the latest statistical tools to detect small changes in small samples with high natural variability. Joanne recalled working like dogs amid the ice statues in Hanover, New Hampshire, to master Bayesian statistics and produce reports and papers analyzing the results of their cloud seeding experiments. She recalled: "I loved our time there even though the temperature was often below minus 30 degrees Fahrenheit and our car rarely started."[28] She learned several important things at Dartmouth. The first was to do

[25] Simpson Papers, 18.10. Simpson's narrative: "My Experience with Weather Modification," March 2006.

[26] Simpson Papers, 2.11. Clipping scrapbook, 1964–1974, Simpson's narrative and clippings re: Joanne and Bob Simpson's work on weather modification experiments.

[27] Simpson Papers, 4.6–4.7. Scrapbook: "Farewell to Experimental Meteorology Laboratory of NOAA Job," 1974, includes farewell letters to Simpson, copies of photographs.

[28] Simpson Papers, 18.10. Simpson's narrative: "My Experience with Weather Modification," March 2006.

Bayesian statistics with a fairly easy program in Basic language on a modest computer. The second was the way to calculate how many cases or data points she would need to reach significance, assuming seeding only made a small percentage increase in total target rainfall. The third was finding the huge progress one could make by assuming the rainfall followed a gamma or lognormal distribution, which turned out to be a close fit discovered long ago by other meteorologists. Significantly, she also learned that she was happier working in an academic atmosphere than in an interface between science and society.

By 1971, with the state of Florida experiencing severe drought conditions, NOAA administrators and the governor of Florida put pressure on Joanne to address the practical question of whether rainfall can be increased by stimulating cumulus mergers. Wilmot Hess, head of the Experimental Research Lab in Boulder and Joanne's supervisor, ordered her team at EML to conduct an "operational" seeding program to relieve a widespread drought. It was called FACE, the Florida Area Cumulus Experiment.[29] Joanne thought that this top-down move was ill considered and called it "a political public relations sham."[30] She felt it was a mistake for a research director to order one of his teams to be directly involved in a politically charged operational program, and her staff was divided on whether or not to proceed in a program in which the results would be difficult if not impossible to verify statistically. Those opposed cited the enormous natural variability of rain over a specific area that would swamp the experimental results. Their calculations actually showed that several hundred cases of seeded versus control cases (chosen randomly without knowledge of the participants) would be needed to show a 10–15% enhancement due to seeding effects. Hess was able to provide resources to conduct only sixty-three experiments—too few to draw any robust conclusions. Moreover, planners, politicians, and the public wanted to see causal, not statistical, results. The message from NOAA to EML was, "we can only give you enough resources to do the project for a limited time, and it had better come out with a positive result, or else."[31]

[29] Simpson Papers, 13.2–13.3. "Florida cumulus, modeling Florida clouds merger ideas, September 1969–May 1974" and "Florida Area Cumulus Experiment (FACE) with dynamic seeding, May 1974–November 1976."

[30] Simpson Papers, 18.10. Simpson's narrative: "My Experience with Weather Modification," March 2006.

[31] Simpson, "Interview," by LeMone.

Joanne really did not want to get involved in applied (or operational) weather modification. Her focus was research, preferably unfettered, as she felt there was still much to learn about interactions between clouds. The last straw came when Hess called her from Boulder on a Friday afternoon to inform her of administrative changes in the program. As of the coming Monday, she was to report to a new and unknown supervisor, Eugene Bollay, a former navy meteorologist and eager supporter of private-sector weather modification. The order from Hess came without discussion, and it was pretty rough on Joanne and her group: "We felt like pawns on a chessboard, being pushed around without consulting us by a person who knew much less atmospheric science than we did."[32] If Joanne refused to comply, she would have had to resign as head of EML and perhaps leave NOAA altogether. She hoped that enough good science would come from the experiments to justify their costs, regardless of the statistical and physical ambiguity.

Joanne and her colleagues knew that silver iodide seeding could not relieve a severe drought, due to a lack of suitable clouds. In such conditions, rainfall often evaporated before reaching the ground. The government wanted to mandate practical results without having developed a comprehensive science plan or validation protocols. Historian Kristine Harper's analysis of weather control as a political football in the hands of state actors follows this pattern.[33] In the 1970s, NOAA scientists could not speak freely or plan their own research, which imposed severe bureaucratic constraints on Joanne's science. For example, equations for calculating the target rainfall had to be specified in advance, before the trials got underway, and scientists could not see the data until the whole experiment was over—that is, until several years later. Statisticians came flocking to FACE, declaring that their experiments really had to be fully double blind. The seed decision envelopes were to be handed to the person who dropped the flares, and they were not supposed to know the decision until the rainfall data had been completely analyzed. Actually, the decision was pretty obvious to the team on the aircraft, due to the behavior of the clouds. On most seeded days, the clouds grew much higher and were more active, and mergers appeared to occur sooner and more often. In the double blind experiments, results were still the same, but the total rain increases were still

---

[32] Simpson Papers, 3.12. Narrative (by Simpson): "The Miami Years, 1967–1974."
[33] Harper, *Make it Rain.*

not statistically significant. Worst of all, NOAA managers would only allow FACE three years to verify, which would not provide more than sixty to eighty experimental days on which a "seed" or "no seed" envelope would be opened. Based on that result, seeding would be done with either dummy flares or real silver iodide flares. Joanne's calculations from the Dartmouth sabbatical indicated that hundreds of cases would be needed to verify a rainfall increase of 15%. There would be no sense in continuing FACE if the question of areal rainfall enhancement could not be resolved within a few years with the resources provided. The project was doomed.

At the height of the controversy, in 1971, Joanne appeared on the NBC *Today* show to discuss the Florida drought and the FACE weather modification experiments. Karen, who was ten at the time, remembered her mother "was wearing a strange wig and outfit."[34] Joanne recalled that the host, Hugh Downs, and his garrulous sidekick, Joe Garagiola, went out of their way to make her feel comfortable, but nevertheless she had reasons to be tense.[35] It seems that NOAA administrator Robert White had instructed her to "lie on camera" about the possible negative effects of increasing the rain in Florida. This was an allusion to the interests of the tomato growers who, fearing crop damage from thunderstorms, did not want more rain. Joanne recalled: "White instructed me to say that I was a scientist, and it was therefore not my concern if some interests were negatively affected." But she resisted, since she did not wish to convey the image of an uncaring or unconcerned scientist. Contrary to White's instructions, she informed Downs that FACE experimenters carefully communicated with the individuals and groups who worked or lived in the targeted area, and would not undertake seeding if there was a significant possibility of substantial damage. In fact, the project had recently suspended operations during the tomato harvest to accommodate the farmers. White was extremely angry with her for this, and held it against her. She wrote later, "He never forgave me. This was a factor, if not the major factor that [convinced me] I had to find another job."[36] FACE produced a number of important research papers, as expected, but the practical

[34] Malkus-Benjamin, "Video Interview," by Lipshultz.
[35] Simpson Papers, 4.6–4.7. Scrapbook: "Farewell to Experimental Meteorology Laboratory of NOAA Job," 1974, includes farewell letters to Simpson, copies of photographs.
[36] Simpson Papers, 3.12. Narrative (by Simpson): "The Miami Years, 1967–1974."

results turned out to be inconclusive. Joanne thought this fueled the critics of the highly controversial field of weather modification, ultimately contributing to its demise. To make matters worse, the enemies of weather modification were more vitriolic and more persuasive than its patrons, claiming that all cloud seeding was ineffective and anyone who disagreed with them was a fraud. Joanne began looking around for positions in atmospheric science to get out of what she regarded as an unsatisfactory situation and increasingly distressing hassles. She was able to do this with the confidential knowledge that her husband Bob planned to retire in three years, in 1974, from his position as director of the NHC.[37]

Bob was also experiencing problems at NOAA, specifically regarding his handling of Hurricane Camille. In 1969 Camille struck the western tip of Cuba as a Category 3 storm, and entered the Gulf of Mexico. Due to previous commitments, Bob found himself without any airborne reconnaissance. NOAA's Satellite Analysis Center experts thought the storm was weakening, but, based on his long experience, Bob believed they were wrong and overruled them. He issued a hurricane watch and began making plans for evacuation of coastal areas, and was ultimately credited with saving thousands of lives. Camille made landfall on the Mississippi coast on the evening of August 17 as a Category 5 hurricane. There were 256 storm-related deaths in the US, with property losses estimated at $1.4 billion, the costliest hurricane to date. The following day, Bob surveyed the extensive damage by helicopter with Vice President Spiro Agnew. and related the story of how the forecast and evacuation orders were made in the absence of good satellite information and available aircraft. Agnew was appalled at the poor satellite information, and even more by the failure to have at least two reconnaissance aircraft ready to go.[38] He clearly believed that the government agencies in charge of protecting life and property had failed. On hearing about this, President Nixon "expressed very great concern" (Joanne says he "blew up") over federal failure to forecast the track and intensity of the hurricane, He immediately had Robert White on the carpet and ripped him up one side and then down the other, and then ordered that immediate steps be taken at the highest levels to improve advance warning of weather disasters. White was severely shaken by the encounter, and

[37] Ibid.
[38] R. H. Simpson, *Hurricane Pioneer*, pp. 111–14; "Agnew Reports to Nixon"; *Hurricane Camille*.

feared for his job. Instead of telling Bob Simpson, "Congratulations for going beyond the line of duty in saving thousands of lives," he made him the focus of his anger, saying "Why did you shoot off your face and make me look bad, you villain?"[39] For years, White held a personal grudge. Bob found that a number of his key personnel at the NHC were refused promotions, and he was denied many of his requests for needed support and upgrades.

Women and Success

In 1972, Joanne was invited to a conference on "Successful Women in the Sciences" convened by the New York Academy of Sciences and sponsored by its sections on anthropology and psychology (Figure 6.2). In her autobiographical and prescriptive talk, "Meteorologist," she sketched her life course and struggles to date before distilling some advice for up-and-coming women in science.[40] "I am 48 years old and

Figure 6.2 Joanne Simpson flying during the Florida Area Cumulus Experiment (FACE) in 1973. Simpson Papers, 455, PD-17. Original in color.

---

[39] Simpson Papers, 3.12. Narrative (by Simpson): "The Miami Years, 1967–1974."
[40] J. Simpson, "Meteorologist."

have devoted thirty years, full time, to my work in meteorology…
I have published eighty research papers and collaborated on two books.
I now serve (one day per month) on an environmental council appointed by
the governor of my state. My husband is the director of an important
weather forecast center, and between us, we have five children, all of
whom are adults except my ten-year old daughter, who is with me only
during the two summer months." After a candid recounting of her
family, educational, and employment history, Joanne turned to the
barriers she faced as a married woman. "The most costly of the many
sex-linked obstacles I have encountered have been the nepotism rules
enforced by employers. My personal and married life and child-raising
have surely suffered for the professional attainments I have reached,
while my career, in tum, has been severely limited by my sex. I am currently not convinced that either the position, rewards or achievement
have been worth the cost," which she said had been "staggering." She
was emotionally, physically, and spiritually exhausted from all the years
of colossal and constant effort and did not think that either the scientific
contribution, the position she had reached, or other rewards were
enough to compensate for the terrible price exacted from her and from
those close to her.

Joanne thought that unless things changed dramatically, the difficulties faced by a woman trying to combine top-level achievement
with successful marriage and motherhood were "close to prohibitive."
She listed three different types of sex-linked problems she had encountered: (1) discrimination simply from being a woman, (2) difficulties
from being a married woman, and (3) difficulties from being a mother.
She experienced gender discrimination as a graduate student when her
department declined to recommend her for any fellowship support
whatsoever, when she felt she was not being taken seriously by both
male and female colleagues, and when she was passed over for employment, and on one occasion, summarily fired. The problems of being a
married woman, however, were incomparably greater. They included
finding work close to where her husband works, complying with social
expectations to put her own career second, reduced opportunities for
upward mobility for both partners, and, even when things seem to be
looking up, avoiding restrictive nepotism rules. Joanne experienced all
of these. She definitely put her career ahead of raising her children. She
was able to afford expensive childcare and boarding schools, but worried that her older sons had been deprived of her full attention. In an

angry outburst, her younger son Steven once charged that she had never been more than a "distant figure" in his life.

She provided four basic recommendations to help younger aspiring women gain more confidence and attain their goals. (1) Exceed the male competition by several orders of magnitude. (2) Cold-bloodedly seek a field so under-populated that employers are "scraping the bottom of the barrel." (3) Forego false pride. Learn to type and to teach even if you must work without pay; cultivate important contacts; never take no for an answer; and develop a thick skin toward being disliked by some, even many, people. (4) Learn to put every minute of your time to good use, in waiting rooms, on streetcars, in boring seminars, and always take work with you, wherever you go. Be prepared for an 80-hour week and love your work, for the sheer joy of doing it.

Joanne pushed for reform and became an advocate for female scientist empowerment and an outspoken critic of the existing power hierarchy within atmospheric sciences. She and Peggy LeMone published a research paper on "Women in Meteorology" that assessed the status of female students, graduates, and professionals in the field. They sent out questionnaires asking women about their experiences with numerous incarnations of sexist behavior from colleagues, supervisors, and mentors and combined those responses with information derived from résumés and employment statistics. A vast majority of women respondents reported notable negative effects of sex discrimination upon their careers, and many reported severe effects. Underemployment was more prevalent than unemployment, as women were systematically underestimated and deemed less knowledgeable than men with the same, or lesser, qualifications. To be competitive, a woman PhD had to outperform her male competitors. Some felt they had been hired as token females, and all married women said they faced nepotism rules. Joanne and Peggy championed equal rights for women and noted incremental progress, but concluded that, "although it is palpably true that overt discrimination has reduced during the past twenty-five years, it is equally true that covert discrimination still exists... nearly all the women we reached who had dropped out of atmospheric science reported that their reason was related in some manner to their sex."[41]

[41] Simpson and LeMone, "Women in Meteorology"; Kundsin, *Women and Success*.

Joanne had experienced gender discrimination in the past and been passed over for promotion. She was a research branch head who had never been promoted. Joanne was fiercely independent and anti-hierarchical. She was known, on occasion, to suspend or subvert the rules. She felt she had made too many enemies in management and had been categorized by her superiors as a temperamental troublemaker.[42] By 1973, Joanne felt that politics had utterly invaded her work area. NOAA managers were heading the experimental effort toward disaster. She thought it necessary to remove herself both from NOAA and from the increasingly ugly controversy. She got out just in time to avoid deepening her depression and to retain her reputation as an honest scientist with high integrity. Rather than simply tendering her resignation and publicly airing the real reasons she had decided to leave, Joanne devised a plan that did not harm her, Bob, or those who remained. She wanted to leave for a better job, or at least one that looked better, preferably in academia.

After several years of searching, in late 1973 both Joanne and Bob had landed apparently solid offers elsewhere. Joanne was promised a senior faculty position in the new environmental science department at the University of Virginia in Charlottesville and an assistant professor slot for Roger Pielke, a cloud modeler working with her at EML. Bob received a verbal offer of a position as a research professor, with office space, departmental involvement, and occasional teaching assignments. Even so, he wanted to give NOAA one more try to see whether he could productively stay on, past age 62, as director of the NHC. They wanted to talk to Robert White frankly one last time. White declined to take a call at his office and told them to telephone him at home that evening. When they did this, he made it clear, without saying it in so many words, that he regarded them as troublemakers and he "would not stand in the way" of their leaving. The next day, they telephoned the University of Virginia and accepted their offers.[43] Joanne was offered an endowed chair as the William W. Corcoran Professor of Environmental Sciences, and Bob agreed to go along as a research professor.

At the end of May 1974, ERL held a send-off party for Joanne, a gala Polynesian luau, at the Royal Biscayne Hotel, organized by her ever-entertaining associate Bill Woodley, complete with skits, songs, a steel

<hr/>

[42] Simpson Papers, 3.12. Narrative (by Simpson): "The Miami Years, 1967–1974."
[43] Ibid.

band, and a limbo competition. Well-wishers brought or sent cards, letters, and mementos. One of the cards is reproduced as the epigraph of this book. Joanne looked back at her decade in NOAA as much more positive than negative. The opportunity to work with bright young scientists at EML and travel widely far outweighed the hurts and frustrations imposed by the administration. She even expressed gratitude to Robert White for the job offer that allowed her to continue her career in meteorology and marry Bob Simpson.

Joanne was alone in 1964 when she joined the weather bureau, seeking to link her life and career to Bob's. He provided the emotional stability and love she so desperately needed, and shared her interests in flying, sailing, tropical meteorology, and life in general. It was fulfilling to be able to reunite with her children and establish, perhaps for the first time in her life, an extended network of family, friends, and colleagues. She was excited, as director of EML, to take on grand intellectual challenges regarding the interactions of clouds, hurricanes, and their environment. She was also excited to be able to weave together the threads of her earlier accomplishments in observational, experimental, and computational meteorology and take on grand societal challenges aimed at saving lives and property—perhaps adding prediction and control to her understanding of the tropical atmosphere. She had turned a corner, but still struggled with depression, migraines, and most of all, deep feelings of guilt that welled up from her checkered past. The work-related stresses imposed on both her and her husband were immense. They were embedded in a massive bureaucracy with insensitive and, at times, mendacious administrators. They were embroiled in controversial projects—Stormfury and FACE—and entangled in controversial issues of weather control and public safety that often conflicted with their scientific judgment. These factors, combined with Bob's impending retirement from the weather service, encouraged Joanne to spend her final four years at NOAA quietly exploring the job market, looking for a way out—a professorship that would restore what she had lost at UCLA and prove acceptable to Bob. The University of Virginia beckoned.

7

University of Virginia

It was the most chauvinistic place I have ever been.

JOANNE SIMPSON

Joanne felt immense relief to be leaving NOAA for new opportunities at the University of Virginia. This was her second chance to be a full professor and she wanted to make the most of it. Fourteen years earlier, her tenure at UCLA had produced turmoil and regret. This time around she expected it would be much better. Her motivations were clear: she was fleeing government bureaucracy with Bob at her side (Figure 7.1). They were leaving NOAA at the right moment for better prospects in Virginia.[1] It did not go as planned.

For their pre-employment interviews, Joanne and Bob traveled to Charlottesville, a genteel pre-Revolutionary War settlement, now a picturesque college town nestled near the foothills of the Blue Ridge Mountains, where they would be trading their access to tropical clouds and a sailboat mooring for southern hospitality and bucolic campus vistas.[2] They met with the dean of arts and sciences, other administrators, and potential new colleagues for lunch and outdoor cocktails at Mitchie's Tavern, an eighteenth-century inn on the access road to Monticello. According to Joanne the visit went smoothly, and the dean, a charming historian, made promises of support to both of them, with the understanding that a delay of several months was needed to effect the transition. In the interim, Joanne and Bob traveled widely on a whirlwind trip that mixed business and pleasure. They visited Israel, where Joanne lectured on weather modification, and they toured the countryside. A World Meteorological Organization conference in Nairobi preceded a weeklong safari in Kenya and Tanzania, then visits

[1] Simpson, "Interview," by LeMone.
[2] Simpson Papers, 18.10. Simpson's narrative: "My Experience with Weather Modification," March 2006; Simpson, "Oral History," by Harper.

First Woman: Joanne Simpson and the Tropical Atmosphere. James Rodger Fleming,
Oxford University Press (2020). © James Rodger Fleming.
DOI: 10.1093/oso/9780198862734.001.0001

Figure 7.1 Joanne and Bob in Charlottesville 1976. Simpson Papers, 455, Photos. PD-19. Original in color.

to the major cities of South Africa—all arranged by their Charlottesville friends Mike Garstang and his spouse Elsabe. The Simpsons flew from Cape Town to Rio de Janeiro, Brazil, in late February, just in time for Carnival and Joanne's quick consultation on water resources.

By the time they arrived on campus, however, the dean had been fired, the provost had retired, and the dean of the graduate school had become president of another university. The only promises the new administrators would honor were those made previously in writing; namely, that Joanne would be an endowed professor in the environmental science department with a research appointment at the Center for Advanced Studies, and her promising young colleague, Roger Pielke, would receive an assistant professorship. A new department of environmental science had been formed recently from the forced merger of the geology and geography departments. It had a very broad purview, with only minimal commitments to meteorology.

Broken Promises

Despite earlier promises, the university had no permanent position for Bob, who worked part-time as a research associate and private consultant. He found the transition, from operational hurricane forecasting to quiet campus life, "hard." Shockingly, his job offer, originally predicated on the availability of soft money, had evaporated. Joanne had to fight "like heck" even to get him an office, and succeeded only because her friend Mike Garstang gave up some of his space. Joanne first met Garstang in Trinidad in 1956. He was a young South African master's degree student, and Joanne took him under her wing. She urged him to study for a PhD in tropical meteorology at Florida State University, and kept in touch over the years as he moved his family to the United States. He became a naturalized citizen and joined the faculty at the University of Virginia.[3] It was Mike who had first nominated the Simpsons for their positions.

Initially, Joanne was excited by the prospect of becoming an endowed professor of environmental science, but her happiness soon faded away.[4] She discovered that the University of Virginia was horribly sexist, the climate for women professors there was oppressive, and interdisciplinary work was not well supported. Her colleagues, especially the older ones, did not want her there. They treated her badly, and resented her involvement in faculty affairs. They clearly made it known that, as a woman, her input was, at best, unwanted. This triggered her recurring and deep-seated depression, for which she again sought treatment. She recalled her disappointment and outrage: "It was the most chauvinistic place I have ever been. People said to me, 'You only got the endowed chair because you're a woman', and people wouldn't sit next to me at the faculty senate meetings. All endowed chairs or full professors had to serve on the Dean's Advisory Committee, and this dean was absolutely horrible."[5] To make matters worse, the new dean was, in Joanne's opinion, a "failed physicist" who thought meteorology was not a "real science," and it should not be taught at Virginia. Joanne was the only woman in the department, and her dean went out of his way to make it clear that the only reason she was even there was because he was

---

[3] Simpson, "Interview," by LeMone.; Simpson, "Oral History," by Harper.

[4] R. H. Simpson, *Hurricane Pioneer*, p. 139.

[5] Simpson, "Oral History," by Harper; R.H. Simpson, *Hurricane Pioneer*, p. 119.

forced by the government to hire a woman—"and it choked him. You could just see from his face."[6]

Joanne's survey article on women in meteorology, written with Peggy LeMone, had just been published, It was very timely, since it addressed the virulent sexism Joanne was now experiencing at the University of Virginia. Joanne became a founding member of the AMS Board on Women and Minorities, tasked with helping working women identify and surmount obstacles such as prejudicial treatment, exploitation, and domination.[7] Bob was very unhappy too. He was never made to feel at home on the campus, and he soon invested his energies in a small private consulting business, Simpson Weather Associates, which employed several UVA professors, including Garstang and Pielke. Joanne served as its chief scientist.

National Hail Research Experiment

Joanne had left NOAA, but she remained active in weather modification research for another five years. She devoted her time to testing her cloud model, adding improved simulations of rain, hail, and snow, working on cumulus mergers using data collected by FACE, and participating in a new initiative, the National Hail Research Experiment (NHRE), to understand the formation, and possible suppression, of hail. In 1974 and 1975 the National Science Foundation spent about half of their meteorological budget on hail suppression projects. Joanne wondered, "Why on Earth would they do this?" The reason was to keep up with the Soviet Union, which was assumed to be "way ahead of us in man-made hail suppression."[8] Soviet scientists used radar echoes to identify the hail-growing part of thunderstorms and then fired artillery shells filled with silver iodide into that area of the cloud to convert all the liquid water to ice. Their theory was that there would be many more hail embryos in a seeded cloud than would occur during the slower process of natural freezing. The multitude of hail embryos would take up and freeze all the cloud's water into ice particles too small to become large enough to fall as hail. Success was not evaluated

---

[6] Simpson, "Oral History," by Harper.

[7] Simpson and LeMone, "Women in Meteorology."

[8] Simpson Papers, 18.10. Simpson's narrative: "My Experience with Weather Modification," March 2006.

in the field by collecting data, but by using historical insurance records to compare damage in so-called "protected" and "unprotected" areas. By this questionable verification process, the Soviets claimed to have virtually eliminated hail damage in several vulnerable areas of the country.

Joanne was a member of an international scientific delegation that visited several of their field sites. Radar evidence seemed to indicate that lots of small ice particles appeared shortly after seeding, at least in some cases. Their lead scientists were well read and internationally known. Most members of the US delegation were convinced, but some thought the results were fraudulent and that the insurance claims had been knowingly doctored. Joanne was cautiously optimistic about trying the Soviet seeding technique in the US, mainly because she wanted to learn more about cloud icing processes.

The visit to the Soviet Union motivated the US to establish the NHRE in 1972, a multi-year program conducted by the National Center for Atmospheric Research (NCAR) in Boulder, Colorado, with funding from the NSF. The program set out to investigate whether hail could be suppressed by methods inspired by (but not identical to) those prac- ticed in the Soviet Union, and to improve understanding of severe hail- storms by observation and analysis. The NHRE resources included eight instrumented aircraft, three radars, two sailplanes, and a mesoscale meteorological network consisting of 420 precipitation measuring sta- tions, twenty-two complete surface stations, and four radiosondes. The NCAR Ice Physics Group collected and analyzed hailstone samples. The staff of the NHRE consulted with leading atmospheric scientists and welcomed visitors from around the world. Joanne, who served on the advisory board, thought the immense scale of the project was an over- reaction, motivated by Cold War perceptions of an ongoing "weather race with the Russians."[9]

Unlike the Soviets, the NHRE did not fire artillery shells into thun- derclouds. Instead, they used old-fashioned aircraft-mounted silver iodide smoke generators. But it was too dangerous to fly airplanes into a thunderstorm, so the experimenters had to hope that somehow the plumes of silver iodide smoke reached the target. This was not at all certain. After two years of trying, the NHRE produced no clear results.

---

[9] Fleming, *Fixing the Sky*, p. 176.

Joanne decried the need to employ outside statisticians for experimental design and evaluation. There were never enough resources to fly enough missions to collect robust statistics, and in actuality, the NHRE aircraft did not fly on placebo days because of the high cost. As had been the case in Stormfury and FACE, the presence of the statisticians contributed to a mistrusting atmosphere. Again, Joanne witnessed her colleagues, ordinarily pleasant and peaceful meteorologists, become hypercritical and polarized when involved in weather modification programs, especially when exposed to the public media. Formerly good friends would leave the scientific method aside and tell newspapers that the experimenters are "pigs feeding at the public trough," and call them "snake oil salesmen," and worse. There was no shortage of ill will, as NCAR scientists resented having to work on practical modification projects, and many bad-mouthed the director, who was also being harassed by the NSF and insulted in the larger scientific community.[10]

Like Joanne, Australian cloud physicist William Swinbank, the first director of the project, sought improved understanding of the physical processes leading to the formation of hail. He soon realized, however, that practical results from the project would not be forthcoming. Nature was too complex and the number of seeding trials was too small to generate statistically significant results. One of his Australian colleagues noted that "Bill kept us sane in the face of organized absurdity and didn't hesitate to pin-prick hypocrisy, which he enjoyed doing."[11] Joanne was of the opinion that the pressures and complexities of his responsibilities led to his demise. Swinbank, a workaholic, died of a heart attack in 1973. Into the breach rushed radar specialist David Atlas, who served as the second director of the NHRE—a responsibility Joanne thought "nearly wrecked his career."[12] After several years of poor and inconclusive results, it became clear that hail suppression methods did not work in Colorado, and might, in some cases, enhance rather than diminish the strength of hailstorms.[13]

[10] Simpson Papers, 14.1. "Hail, November 1974 September 1975"; Simpson Papers, 18.10. Simpson's narrative: "My Experience with Weather Modification," March 2006.

[11] Priestly, "William Christopher Swinbank 1913–1973."

[12] Simpson Papers, 18.10. Simpson's narrative: "My Experience with Weather Modification," March 2006; Atlas, *Reflections*, pp. 72–9.

[13] Hitchfeld, "National Hail Research Experiment."

GATE: The GARP Atlantic
Tropical Experiment

In August and September 1974 Joanne flew to Dakar, Senegal, West Africa to participate in the Global Atmospheric Research Program (GARP) Atlantic Tropical Experiment (GATE), led by the World Meteorological Organization and the International Council of Scientific Unions.[14] GATE—perhaps the first double acronym in meteorology— was going on in the Atlantic Ocean west of Africa. GATE involved more than twenty nations pooling their talents, expertise, and resources to examine the key roadblocks in global weather prediction and related problems involving the air–sea interface and cloud and storm systems. Its immediate goal was to examine the ways tropical clouds and smaller-scale weather systems influence larger-scale circulations, with the aim of parameterizing the interactions and incorporating them into numerical models.[15] Its long-term goals included understanding the predictability of the atmosphere, improving numerical models of the global circulation, and extending the time range of daily weather fore-casts to more than two weeks. GATE constituted the initial step toward these huge goals, which were right up Joanne's alley. She was tutored at Chicago in the big-picture general circulation science of Rossby and Starr, she had launched her career at Woods Hole conducting observa-tional campaigns to discover how clouds and smaller-scale phenomena interact with the larger circulation, and she had developed a computer-ized cloud model that began to take into account the microphysics of water vapor, cloud water, and ice. The scale and importance of GATE was a welcome relief from the local turbulence going on in Charlottesville.

The GATE observational program had four scales ranging from very large to very small. The experimental area comprising the A-scale extended from the westernmost part of the Indian Ocean, across trop-ical Africa, the Atlantic, and South and Central America, to the east-ernmost part of the Pacific Ocean—about 40% of Earth's tropical belt between 20° N and 10° S. The B-scale was defined by a concentrated array of ships in the eastern Atlantic, arranged in a hexagonal pattern about 1.5° apart and designed to study the life cycles, bulk properties, and environment of cloud clusters. The C-scale used vertically stacked

---

[14] "GATE, History of the GARP Atlantic Tropical Experiment."
[15] Kuettner, "General Description and Central Program of GATE."

aircraft, supported by five ships stationed 50–100 km apart, and a dense network of buoys. It was designed to examine the structure of different types of convective disturbances, their life cycles, and the vertical and horizontal fluxes of mass, heat, moisture, and momentum. In the D-scale, individual aircraft flew into convective "hot towers" to measure vertical motions, liquid water content, and other cloud physics parameters. This system of telescoping scales comprised a nested grid. Over and above it all, NASA's Synchronous Meteorological Satellite, SMS-1, launched in May 1974, provided visible and infrared images with resolutions of 0.5 and 5 nautical miles, respectively. Three US and one Soviet polar-orbiting satellites provided additional resources. Joanne, who had never had such space-based support, called the satellite imagery "beautiful." It was an experience that helped shape her next major career move.

Joanne found the field program exhilarating. She learned an enormous amount and made friends and colleagues from all over the world. There was no need to search for clouds, since the experimental area was one of the cloudiest oceanic areas in the global tropics. She flew on all thirteen of the instrumented aircraft in GATE, and she particularly enjoyed flying on the Soviet aircraft, where the scientists were still making calculations on slide rules (as she had done in the mid-1950s), while they greatly envied her HP 35 pocket calculator.

Joanne was the mission scientist for day 261. Harkening back to her experiences on the PBY at Woods Hole and in the Pacific Cloud Hunt, she arranged all the aircraft tracks to map clouds using photogrammetry. The aircraft flew a square pattern 80 km on a side at a number of different levels. The key aircraft all had side-looking cameras on both sides. Using the time it took a cloud feature to cross the screen when projected, the distance and height of each cloud feature could be obtained. Joanne and her team later, and laboriously, constructed a map of virtually every single cloud within the array.[16]

Joanne was generous in her praise. Immediately after the field phase she wrote that GATE had "more than achieved its goals and had exceeded her own and most scientists' expectations." She credited the cooperation of the dedicated people, groups, and nations involved, clear objectives, effective management, and "just plain good luck." Never in her thirty years in the profession had she seen so many untried systems work in their first major field program. "Young and old complemented and stimulated each other. Famous scientists and graduate students together sweated [through] 15 to 20-hour mosquito-infested days, broiled and froze on aircraft . . . , and performed good old manual labor, which no one fancied himself or herself too good to do, since it had to be done."[17] She did not get the impression, as in other examples of "big science," that politics was dominating, nor that there were too many cooks in the kitchen. Joanne admitted to having at least one temper flare-up, however, triggered by exhaustion, probably as a result of flying for two days straight with only one day of rest.

GATE intrigued Joanne, since the observations revealed many new phenomena she referred to as "puzzles." She credited her colleague Robert Houze for the new insight that convective clouds produce significant amounts of anvil and stratiform precipitation, in amounts comparable to convective precipitation. She credited her colleague Ed Zipser for insights about two main types of cloud systems: slow-moving clusters and fast-moving squall lines that can outpace the strongest wind. She credited Mike Garstang for showing her the importance of dry downdrafts that shut off convection for up to eight hours. She worried, however, that the GATE region might not be representative of the tropics around the world. "My pre-GATE contention that the GATE area is almost a 'freak' region of the tropics is stronger than ever as a result of my participation."[18] She was "highly suspicious" of generalizing to the global tropics any parameterization schemes developed and tested in the GATE area. She speculated that there were as many as four to five types of tropical regions that needed to be treated separately: Atlantic (eastern and western); Pacific (eastern and western); and continents. Finally, she predicted that cumulus modelers might be in for rough going in their attempts to simulate the GATE cumuli, since they were more transient than those in Florida or the Caribbean and had poor, uneven, and variable bases.

She praised the forecasts for convection in the B-scale as "the best job of tropical forecasting (omitting severe disturbances) that I have ever had the pleasure to observe." She attributed it in part to luck, but also credited the satellite imagery. On the whole, the plusses in GATE exceeded the minuses by an order of magnitude. Virtually everything

---

[17] Simpson, "GATE Aircraft Program."
[18] Ibid.

was done either well, or the best it could be done under the circumstances. Dozens of groups dedicated themselves unselfishly to the achievement of the GATE goals. These goals were, in fact, more than achieved in the field phase.

Joanne thought that GATE overemphasized large-scale modeling processes—parameterizing convection and scale interactions, when her focus was on the small scale—cloud physics, and microphysics, which were not well understood. She thought that the fundamental assumption of global modelers—that the "important" effects of cumulus convection can be incorporated into the larger-scale models without the detailed knowledge of individual clouds and their microphysics— was true to some extent, but stood in need of refinement. Her experiments in Florida had demonstrated the critical importance of such often-neglected parameters as vertical distribution of cloud water content, evaporation of precipitation, and local wind shear.[19]

GATE stimulated work on a number of improved two- and three-dimensional cloud models for mesocyclones, tornadoes, and waterspouts. Gary Van Helvoirt—one of Joanne's graduate students at Virginia—used a model developed for large weather systems over the US Great Plains to simulate GATE clouds, but the model overestimated updraft and downdraft velocities. The model revealed, however, that cloud motions within a mesocyclone were rotating around a vertical axis and could spin out the waterspouts observed in GATE.[20] What roles do clouds play in the tropical atmosphere? How do the cumuli, which occupy such a small fraction of the area, distribute their concentrated heat, moisture, mass, and momentum to the synoptic scale? Hearkening back to her thesis at Chicago, GATE allowed Joanne to make direct measurements of physical and microphysical cloud properties to compare with the synoptic models. Much of her research during her five years at Virginia involved working up the results of the GATE field experiments and reporting on them with Claude Ronne (Figure 7.2). Her papers include mission notes, photographic logs, conference papers, and about a dozen GATE-related publications on clouds, cloud models, and waterspouts.[21]

[19] Simpson and Wiggert, "Models of Precipitating Cumulus Towers."
[20] Simpson Papers 3.13. Projects: "Mapping clouds from GATE," 1971, 1974, includes material from Simpson's work at University of Virginia and Simpson's narrative on work.; Sherer, "Who's Who in GATE."
[21] Simpson Papers, 13.4–15.2; 15.4, 15.6. Materials relating to the GARP Atlantic Tropical Experiment (GATE).

Joanne was concerned about GATE's legacy and whether resources would be available for subsequent analysis of all the data. Previous massive projects, such as the International Geophysical Year, had been left unfinished, in part due to lack of funding and the lack of glamour involved in the reanalysis of data. She estimated that proper field program analysis required five to ten times the financial expenditures as the actual operation, and twenty-five to fifty times the time expended— an investment for the future that could be optimized by involving more students. She wanted the younger, less known, less well-funded meteorologists to have access to the data. She felt that the GATE data, if followed up and analyzed properly, would advance both tropical and global meteorology by a significant amount. As the twenty-fifth anniversary of GATE approached, Joanne rated it as "the best field program of my lifetime," adding that it generated the most useful data set in tropical meteorology and resulted in the most significant publications.

Joanne also participated in MONEX—the Monsoon Experiment, part of the international GARP effort to achieve a better understanding of the onset and behavior of monsoonal circulations. The goals of the project were planetary and practical, involving the major seasonal perturbation of the general circulation of the atmosphere and the influence of the annual cycle of precipitation associated with monsoons on the agriculture and well-being of the many populous nations of the region. MONEX observational studies took place during the winter and summer monsoon seasons of 1978–79 over Southeast Asia and the South China Sea, India, the Bay of Bengal, Western Africa, and the Arabian Sea.[22] It was in effect, an attempt to apply the techniques developed for GATE in the tropical Atlantic to the tropical landmasses of the world.

GATE and MONEX were mainstream atmospheric science projects, but Joanne also explored some peripheral and somewhat quirky issues. In 1977 she was invited to an international conference in Iowa on "Iceberg Utilization," funded by Saudi Arabian Prince M. Al-Faisal. The conference announced the coming "Iceberg Age" and explored visionary ideas about using icebergs as alternative water and energy resources relevant to the growing concerns about global water shortage and energy availability. In a keynote address, prominent journalist Lowell Ponte raised concerns about possible adverse environmental effects of

---

[22] Murakami, "Scientific Objectives of the Monsoon Experiment (MONEX)."

Figure 7.2 Joanne, Claude Ronne, and Bill Snow working on GARP data at the University of Virginia, 1975. Joanne's handwritten caption: "Information about the people working. Photographer: Claude Ronne, who also flew with me in the Pacific flights. Note the stereoscope to the right. Man with beard: Graduate student Bill Snow. He is operating the large 35-mm motion picture projector. The clouds were mapped by frame counting, following each feature across a gridded screen." Simpson Papers, 455, Photos. PD-18. Original in color.

towing icebergs, including inadvertent weather and climate modification. In response to eager and earnest desires to bring water to a thirsty world, Ponte pointed out the irony of taking ice *from* a desert (Antarctica) and bringing it *to* deserts (in the Middle East, for example). Joanne's paper compared the costs and efficacy of towing icebergs to lower latitudes with cloud seeding, and she examined potential atmospheric impacts caused by cooling and melting of fresh water.[23] The paper was highly speculative, but concluded what was already known—that the water provided by one large iceberg would greatly exceed the additional rain that cloud seeding might produce in arid areas. She also published

[23] Simpson Papers, 15.3. "Iceberg [conference] talk; Simpson, "Iceberg Utilization"; Ponte, "Alien Ice."

a paper on weather modification in Virginia, but this work threatened to reopen old political controversies, this time farther north, in the mid-Atlantic region. These publications were not widely read and were not part of any larger scientific efforts.[24]

While on the faculty at Virginia, Joanne served the AMS as a member of its board on women and minorities, and committees on cloud physics, hurricanes, and tropical meteorology. She was elected to Council in 1975 and to the Executive Committee in 1977. The National Academy tapped her for both its GATE and GARP advisory panels. She joined the boards of the NHRE and the *Virginia Journal of Science*, and was an associate editor of *Reviews of Geophysics and Space Physics* and a consulting editor for *Weatherwise*. The Weather Modification Association presented her with its Vincent J. Schaefer Award in 1979, and *Ladies Home Journal* nominated her for Woman of the Year in Science for three consecutive years.

Joanne had occasions to visit with and attempt to reconcile with her mother, Virginia, who had moved into a high-rise apartment in a retirement community that Bob had found for her in Virginia Beach, some three hours away. She enjoyed the ocean view and, for a number of years, invited Bob and Joanne down to share Thanksgiving dinner with her and her friends. Joanne hoped their relationship was improving, but their interactions remained stilted. She winced each time they visited when she saw the Benjamin Russell whaling painting hanging above the fireplace, which Virginia had appropriated at the time of her divorce. Seeing the painting never failed to irritate Joanne, who felt that her father had promised it to her. Virginia knew that Joanne cherished it, but made sure to mention on each visit that it was destined to go to her brother Dan. The painting served as a constant reminder of all the problems of the past, all the niggling insults of the present, and most of all, the devastating lack of a mother's love. Joanne eventually purchased an early draft of the painting in an art shop in Nantucket, but said the loss of the original picture was a source of deep regret to her. Joanne's step-daughter, Lynn Gramzow, witnessed first-hand the shocking difference in Virginia when she was interacting with Joanne, compared to her interactions with friends and acquaintances. Virginia stopped her Thanksgiving invitations sometime before 1985, but

---

[24] Simpson and Brown, "Potential of Summer Rain Augmentation."

continued to invite other family members. Joanne kept trying to please her mother, but it was a losing battle.[25]

Moving On

Bob and Joanne soldiered on at the University of Virginia, but, for obvious reasons, they never felt really happy there. They were outgoing and effective mentors for a cadre of young scholars, and their professional services were in demand, but their circle of friends and acquaintances was quite small. They were seriously considering moving elsewhere. In the summer of 1978, Joanne gave NATO training lectures in London and was able to renew old acquaintances. The Simpsons visited Scotland for a week accompanied by Bob's daughter Lynn. Ruing an imminent return to the constricting atmosphere of Charlottesville, Joanne and Bob took walks together in the highlands and came to the conclusion that they had to seek another venue as an outlet for their professional pursuits.[26] The final straw came in 1979. The environmental science department was searching for a professor of paleoclimatology. Joan Feynman, whose brother Richard was a celebrated physicist, visited and gave an excellent seminar that excited the students, faculty, and even the "grumpy" chair of the department. When the department brought a request to hire her to the dean, he pointed at Joanne and said "We are required by the federal government to have 0.2% of our arts and science faculty women, and *you* are it." He would not grant Feynman an interview, and declined any further consideration.[27] Joanne was never angrier in her life, but had learned, from the school of hard knocks, to suppress her outbursts. Then and there she decided, "I'm leaving."

Joanne had heard that David Atlas, one of the pioneers of radar meteorology and the second director of the NHRE, had gone to NASA. He was building a new laboratory in atmospheric science, a kind of NCAR-East, and was looking for people. She had known Atlas for a long time, and that evening she called him up and said, "Hey Dave, I hear you have a great laboratory. Do you need any people?" And he

---

[25] Simpson Papers, 1.15. Family history: Simpson's narrative: "My Mother and I," January 1998, documenting Simpson's difficult relationship with her mother and captions for family photographs from the 1930s to the 1980s.

[26] R.H. Simpson, *Hurricane Pioneer*, p. 119.

[27] Hirschberg, "My Mother, the Scientist" [on Joan Feynman].

said, "When can you come?" He said he had been searching for two years to find somebody to be the head of the Severe Storms Branch, and added, "If I had thought that you would even consider coming, I would have asked you before."[28] Within a very short time, the Simpsons moved to Maryland. Joanne took a one-year leave of absence from the University of Virginia, and then extended it to two years. She had always been running scared regarding employment, so she had a policy of never burning bridges. After the second year, the University said "Either come back or resign." To tie up loose ends, Joanne went back to Charlottesville for one day and talked to the current department chairman and a few others. At the end of that day she said, "I just can't come back here. I'd rather be dead." And so she moved permanently to NASA's Goddard Space Flight Center.[29]

[28] Simpson, "Oral History," by Harper.
[29] Simpson Papers, 4.8. Simpson Symposium, February 9–13, 2003; Simpson, "Oral History," by Harper.

8

NASA

I've looked at clouds from both sides now.

JONI MITCHELL

Greener Pastures

Joanne Simpson left the University of Virginia in 1979 to become head of the Severe Storms Branch of NASA's Goddard Laboratory for Atmospheres (GLAS), an elite center for atmospheric science that emphasized space-based remote sensing and computer modeling. Her direct supervisor, radar specialist David Atlas, had arrived at NASA two years earlier to build a research program in meteorology. He referred to Joanne as the "intellectual center of gravity" for the severe storms program, and praised her ability to attract other first-rate scientists, lead "a magnificently imaginative and productive" research agenda, and acquire the necessary resources from NASA headquarters.[1] This was Joanne's last professional move, and it was a good one. Her signature efforts involved improving cloud models and serving as the study scientist and then project scientist for the new satellite program, the Tropical Rainfall Measuring Mission (TRMM). The goals of the program included using radar to measure tropical clouds and rainfall from space, and quantifying their effects on the hydrological cycle, the general circulation of the atmosphere, and Earth's climate. She had looked up at the clouds all her life; she had flown under, over, around, and through them; she had modeled them on paper and with a digital computer; now she would be able to look down on them from sensors in outer space.[2]

Joanne worked at NASA until her retirement in 2004 at age 81. She found professional fulfillment there as the chief scientist for meteorology and a senior fellow, and garnered top awards and multiple honors, many of them firsts for women. She found peace of mind and

---

[1] Atlas, *Reflections*, p. 89.
[2] Kummerow et al., "(TRMM) Sensor Package."

First Woman: Joanne Simpson and the Tropical Atmosphere. James Rodger Fleming,
Oxford University Press (2020). © James Rodger Fleming.
DOI: 10.1093/oso/9780198862734.001.0001

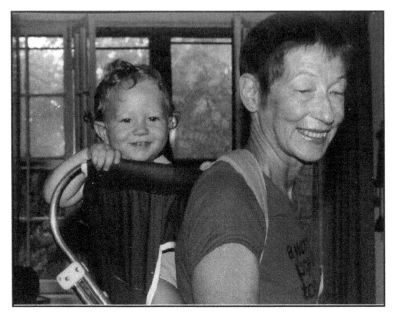

Figure 8.1 Joanne with grandson Christopher, age 2, in 1982. Christopher was the first of six grandchildren. Simpson Papers, 455, Photos. PD-15. Original in color.

stability with Bob and their blended family, which now included grandchildren (Figure 8.1). She was also able to resolve many long-lasting tensions regarding her relationship with her mother, Virginia.

Joanne joined an exciting laboratory at Goddard with highly competent and interesting colleagues. She also joined a community where being a woman made no difference to the way you were treated. The second day on the job, she encountered two other women scientists discussing meteorology in the ladies' room. She recalled, "I had never in my life been at a place before where anyone else other than the secretaries and me used the ladies' room."[3] In this environment, Joanne was able to build and support a strong team of both women and men, and they responded with utmost loyalty. She said of NASA Goddard, "It's sort of like a small town and it has a really nice atmosphere. I mean there aren't cliques hating each other, which was the way it was at the University of Virginia."[4]

[3] Weier, "Joanne Simpson (1923–2010)"; Raymond and Carlson, "My Daughter the Scientist."

[4] Simpson, "Oral History," by Harper.

Her new associates included Robert Adler, then head of the Mesoscale Processes Branch, who had earned his doctorate at Colorado State University, and Louis Uccellini, a postdoctoral specialist in severe storms out of the University of Wisconsin and one of her key mentees at Goddard. Uccellini was a prime witness to the changes she brought:

> Clearly she was a tremendous leader and a tremendous business manager in dealing with NASA headquarters and the managers within Goddard. Before she arrived, we were fibrillating as a unit, the reason being that we were getting bounced around between urgent orders coming from headquarters, and when she came on board, she filtered all of that out. She gave us a straight signal: we're going to develop a strategic plan. This is what we're going to do over the next five years. She fought for that, and all the ups and downs we were seeing beforehand disappeared. I called her "the best band pass filter" I've ever seen in my life. So it was really tremendous, and I saw the value when she did that, and she made it work for us below her.[5]

NASA was trying hard to be inclusive in providing top opportunities for women. In 1978, the Space Shuttle astronaut and mission specialist program announced that it had added six female trainees. The US was playing catch-up with the USSR. Soviet cosmonaut Valentina Tereshkova had orbited Earth in 1963, but NASA's female astronaut corps of that era never got off the ground. At a press briefing, Catherine Sullivan, a spokesperson and one of the six, emphasized how important it was in that group to have the support of cohorts when dealing with NASA's all-boys network. Uccellini, who was sitting in the front row when he heard those remarks, immediately thought, "You know, Joanne had to do it alone . . . Joanne did it all alone, when she was doing these flights; she was the only woman on the plane. You know, when she was trying to organize experiments, trying to deal with Rossby at the University of Chicago, she was the only one."[6]

At the end of her second year of absence, the University of Virginia asked Joanne to return or resign. She decided that the interesting work and congenial colleagues at Goddard were more important than a fancy job title and higher salary at the university. She called the decision to stay at Goddard, the best one of her professional life.

Not everything went smoothly, however. During the presidential election campaign of 1980, candidate Ronald Reagan promised a federal hiring freeze and instituted it the day after he was inaugurated. This

[5] Uccellini, "Interview," by Fleming and Meltsner.
[6] Ibid.

initiated a scramble at NASA to hire new staff before the deadline. Joanne lined up the paperwork to add three new members to her group and sent Uccellini on the run to get the necessary signatures. As he dashed back across the Goddard campus, he stepped into a hole and sprained his knee. He limped into Joanne's office, but she was not there. The following day, Joanne was back in the office with a cast on her arm, and he was on crutches. She was so excited about the hiring that she tripped over the telephone line and broke her arm. Uccellini cited this as an example of her high level of enthusiasm. "She really did go all out, and really was supportive of everything we did. She was a leader. We wrote the proposals, and she built the team."[7] David Atlas remained head of GLAS until 1984, when he was forced out due to conflicts with program managers at NASA headquarters. He had not made friends downtown when he called them "the abominable no-men." Atlas was succeeded by Marvin Geller, whom Joanne called, "the best boss I ever had . . . essential for the existence of TRMM."[8] Honors accrued quickly. In 1982 Joanne received the NASA Award for Exceptional Scientific Achievement, and in 1983 the most prestigious award of the AMS, the Carl-Gustaf Rossby Research Medal (Figure 8.2).

Remembering Claude

In November 1984, while visiting Australia, Joanne received the unwelcome news that Claude Ronne, her first real love and long-time colleague, had passed away. Claude was at the center of her life at Woods Hole in the 1950s, and was centrally involved in the turmoil surrounding her move to UCLA in 1960 (Chapter 5). Joanne had flown with Claude in her NOAA years (Chapter 6), and worked with him on the GATE imagery at Virginia (Chapter 7). She kept the GATE materials in a drawer at Goddard. They had maintained a platonic relationship in later life, and with Bob's approval she visited him at Cape Cod every year. Joanne, who knew Claude was terminally ill, wrote in her private journal after his death, "and I was not there."

> Your death may be final for you—it can't be for me; you will never be far from me; I carry on conversations with you in my mind as always. When I first started writing in these notebooks as a very young woman in my

    [7] Ibid.
    [8] Simpson Papers, 4.8. Simpson Symposium, February 9–13, 2003; Simpson, "Interview," by LeMone.

Figure 8.2 Joanne with the Carl-Gustaf Rossby Research Medal, presented to her in 1983 "for her outstanding contributions to . . . understanding convective clouds and the role of convection in the formation and maintenance of hurricanes and other wind systems over tropical oceans." ("Awards," American Meteorological Society.) Simpson Papers, 455, Photos. III-53. Original in color.

late twenties, I was so afraid of our relation ending in estrangement. I guess I am surprised, and we were lucky, that it took the grim reaper to end it—but he did not. I carry you in my heart and mind with me, as I have almost since I met you in 1947 and as I did during the 24 years that passed since my last entry in this notebook. Twenty-four years is a whole generation. My sons grew from babies to middle-aged men; my little Karen appeared and is now a young lady of 23—yet we went on, we changed, we lived through traumas . . . I was with you in love until nearly the last. My life has been transformed and my vision larger, you will be with me, part of me, so long as consciousness remains . . . So today it is not only *Resquiescat in Pace*, my dear one, the jewel of Cape Cod, but also to a young passionate woman who moved in the sun with you, in P'town, in Onslow Gardens, Kensington, on Anegada, Montego Bay, in Falmouth with the whippoorwill at dawn so many times.[9]

[9] Simpson Papers, 1.7. Personal Diary III: 346–7, 360, Sept. 29, 1985.

From 1985 through 1995, Joanne and several other women were featured in an exhibit on women in science that toured the country. It was created by the Chicago Museum of Science and was called "My Daughter the Scientist." Fortunately, the organizers dropped the original title, "From Petticoats to Lab Coats." Most of the women featured were selected from the 1974 book, *Successful Women in the Sciences*, the result of a New York Academy of Sciences workshop on professional women. Joanne included a model of the instrumented aircraft she had used at NOAA for experiments on clouds, a roll of cloud photographs from the 1957 Pacific Cloud Hunt, a dummy aluminum silver iodide flare used in cloud seeding, and a model cloud constructed of vertically mounted transparency sheets depicting the genesis of a waterspout (Figure 8.3). She added her ballet shoes and a picture of King Richard III to signal her interest in British history. She said the exhibit organizers wanted to portray the scientists as human beings with feelings and frailties, rather than the stereotypical image of a

Figure 8.3 Joanne and Karen *c*.1985, with the exhibit "My Daughter the Scientist." Simpson Papers, 455, Photos. PD-20. Original in color.

white-coated scientist peering at test tubes through horn-rimmed glasses, writing down world-shaking numbers and equations. Joanne designed the exhibit in the year that Claude died. It was not entirely a coincidence that her section of it featured so much of what she and Claude did together, including their discussion about Shakespeare's Richard III on the day they first met. Oversize boxes in the Simpson Papers contain several of the artifacts from the collection, including the roll of cloud photographs, the model cloud, a dummy cloud-seeding flare, and her ballet shoes.[10]

In 1995, Joanne's love for Claude was still strong. She wrote about her own experience to her son David, who was going through a divorce and had a new romantic partner: "From 1947–1984 I had two men in my life, and one of them was Claude, and that love not only lasted until he died. It is still there, and I miss him terribly 12 years after he died. That does not detract from my love for Bob, who is my life's companion. Claude would not have been strong enough to take on the children, my work, and my emotional ups and downs, and fortunately, we were both smart enough to see that."[11]

Tropical Rainfall Measuring Mission

In September 1984, Goddard scientists Gerald North, Thomas Wilheit, and Otto Thiele submitted a proposal entitled "Tropical Rain Measuring Mission." It had been more than a decade since the NASA weather and climate community had won approval for a new research mission. TRMM would carry a number of instruments, including a microwave radiometer, a visible/infrared radiometer, and the first weather radar to be flown in space. The project was inspired by Joanne's earlier work in measuring and modeling clouds and her publications with Herb Riehl on the tropical heat budget.[12] The satellite would fly in a low-altitude orbit (about 320 km) to provide high-resolution images and information from a poorly understood and poorly monitored, but crucially important region of the world. It would be a "flying rain gauge," operating between latitudes 37° N and 37° S, where more than two-thirds of

[10] Simpson Papers, 3.4 and Oversize boxes 19F+B.1m, 2m, 3, and 4 m.
[11] Simpson Papers, 1.3. Correspondence of Simpson with son David, 1995–1996.
[12] Riehl and Malkus, "On the Heat Balance in the Equatorial Trough Zone"; Riehl and Simpson, "Heat Balance of the Equatorial Trough Zone, Revisited."

the global precipitation falls.[13] These measurements, the first of their kind, would provide data on the relationships between rainfall variability, latent heat release, and short-term climate changes. The proposal received funding, and the project moved forward.

In 1986, NASA asked Joanne to be the project scientist for TRMM. She brought to the effort four decades of experience in tropical meteorology, fundamental and hard-won knowledge of convective clouds and tropical storms, and a very strong interest in precipitation and its impact on climate processes. From that time until the satellite launch in November 1997, she worked in close partnership with NASA headquarters staff, the project engineers (whom she admired greatly), and the scientists she recruited to develop the data system. As project scientist, she led the development of the TRMM science goals and the observational requirements necessary to achieve them. Undoubtedly, the community's experience in the GATE program sowed the seeds of the TRMM concept by demonstrating the feasibility of satellite-based radar.[14] In particular, when researchers compared data from shipboard radar with data from the spectrograph and passive radiometers on Nimbus 5, they realized they could also fly radar in space to measure rainfall over the global tropics.[15] Joanne and her team conducted calibration and feasibility studies, flew airborne versions of the proposed remote sensing instruments, and compared them with ground-based observations of rainfall.

Joanne was politically astute in building support for the project from NASA headquarters, from members of Congress, from scientific colleagues, and from the general public. She and her coworkers shared TRMM science and humanitarian goals with meteorologists in a widely read 1988 article.[16] "Precipitation is the most crucial link in both the hydrological cycle and the global atmospheric energy budget. It provides life-giving fresh water resources and produces life-threatening storms and floods. The latent heat released by tropical precipitation helps drive low-latitude circulations and supplies energy to power the

[13] Simpson, "Satellite Mission to Measure Tropical Rainfall"; Simpson et al., "On the Tropical Rainfall Measuring Mission"; Theon, "My View of the Early History of TRMM"; Conway, *Atmospheric Science at NASA.*

[14] North, "GATE and TRMM."

[15] Simpson Papers, 2.7. Awards: International Meteorological Organization Prize, 2002, includes press release, Simpson's narrative, acceptance speech, event program.

[16] Simpson et al., "Proposed Tropical Rainfall Measuring Mission (TRMM) Satellite."

wind systems and balance the global heat budget. Yet rainfall had remained one of the most difficult of all atmospheric variables to measure, especially over the oceans."[17]

On the occasion of receiving a distinguished alumna award from her beloved Buckingham School, she explained the TRMM satellite in plain English: "Right now at NASA, I am serving as the lead scientist in getting a new satellite designed, built, and launched in 1997. After that, we will analyze the first accurate measurements of rainfall over the tropical oceans. The important contribution of rain and its variations to Earth's climate system is clear, but now it is not known within a factor of two, making global warming calculations quite risky."[18]

Although approved in principle, financial considerations kept the program in limbo for several more years. In 1990 Joanne was instrumental in gaining Congressional approval and giving TRMM a new start as a joint NASA–Japanese National Space Development Agency (NASDA) program.[19] The Japanese joined the proposal as an international partner. Nobuyoshi Fugono managed the program, and Ken'ichi Okamoto led the team designing and building the precipitation radar.[20] NASA was responsible for building the satellite and the other instruments, including a microwave imager, visible and infrared scanner, lightning-imaging sensor, and CERES, the Clouds and the Earth's Radiant Energy Sensor, intended to measure cloud amount, height, thickness, and particle sizes. The Japanese agreed to launch TRMM on their new H-II rocket.

As Joanne recalled, "the number of things that could have gone wrong were absolutely so many that it was frightening." But all went well. The budget and the schedule were tight, but the launch from Tanegashima Island, Japan, on November 27, 1997, succeeded, and all of TRMM's antennas and instruments deployed. The data acquisition system, designed by Chris Kummerow, was ready a year ahead of time. At $75 million, TRMM was one of the few NASA projects that came in on time and under budget, with all its instruments working.[21] Several Goddard employees said that "It changed the culture

[17]   Ibid.

[18]   Simpson Papers 2.6. Awards: "Milestones and Awards, 1980s to 1990s."

[19]   Simpson and Kummerow, "Tropical Rainfall Measuring Mission."

[20]   Simpson Papers, 4.8. Simpson Symposium, February 9–13, 2003; Tao and Adler, eds., *Cloud Systems, Hurricanes, and the Tropical Rainfall Measuring Mission (TRMM)*.

[21]   Simpson, "Oral History," by Harper.

of NASA." The satellite had fuel for a 4–5-year mission, but ended up working far longer, splashing down in 2015. After the successful launch of TRMM, Joanne stepped down as the project scientist and, with the new data, including the first images of radar echoes from space, focused on tropical cyclone research (Figure 8.4).

Later in life she referred to her involvement with TRMM as her greatest accomplishment.[22] Her career and the success of TRMM ran in parallel, but, as had been the case for so many decades, it was her models that connected observations and theory. Joanne preferred to work with a cloud modeler rather than a large-scale or global modeler. She was able to recruit W.-K. Tao, a postdoctoral scholar from the University of Illinois, who brought a cloud ensemble model with him to Goddard. Although technically assigned to another division, Tao reported directly to Joanne.[23] They worked on a realistic model of cloud mergers and ensembles. It was important to link theory with observation and to calculate derived quantities.

Figure 8.4 Three-dimensional TRMM radar image of precipitation in super-typhoon Bopha (Pablo) taken on December 3, 2012, just prior to making landfall on the Philippine island of Mindanao. Hot tower convection reached altitudes of over 16 km. Courtesy of NASA/SSAI, Hal Pierce. Color image available at https://www.nasa.gov/images/content/710839main_20121203_Bopha-TRMM_full.jpg

[22] Weier, "Joanne Simpson (1923–2010)."
[23] Uccellini, "Interview," by Fleming and Meltsner; Sherer, "Who's Who in GATE."

By 2002, TRMM had met or exceeded nearly all of its important scientific goals. One, however, still remained: to estimate the profile of latent heating released by tropical cloud systems. The difficulty had always been that latent heat cannot be measured directly. TRMM was now providing data that allowed new cloud models to estimate latent heat release in the tropics. Her work in the 1950s with Riehl on trade wind convergence and latent heating of the atmosphere had come full circle. With the help of her modeling group at NASA and her former colleagues elsewhere, a new multi-scale modeling framework had replaced conventional cloud parameterizations by embedding a cloud-resolving model into a larger general circulation model, allowing for two-way interactions between the models and explicit simulation of smaller-scale cloud processes as well as global coverage. The Regional Atmospheric Modeling System (RAMS), developed by Joanne's former mentees Bill Cotton and Roger Pielke at Colorado State University, links models of clouds and local circulations with models at a much larger scale.[24] Climate modeler Warren Washington recalled how Joanne's insights provided a "path forward" to include realistic clouds and convection processes in global climate models.[25]

American Meteorological Society

In 1989, Joanne was elected President of AMS—the first woman to be so honored. The Simpson Symposium Scrapbook (Simpson Papers 4.8) includes a low-resolution snapshot of Joanne flanked by 20 of her male colleagues who were serving in leadership roles. Joanne was alone at the helm. In her report to the membership in 1990, just a few months after the fall of the Berlin Wall, she pointed out that while "peace had broken out," a new challenge had come to the fore.[26] Global environmental change had become a major concern for humankind, bringing with it various implications, challenges, and opportunities for meteorology and for education.[27] AMS was squarely in the spotlight as never before, at the interface of public information and policy initiatives. She expanded on this in her presidential remarks: "World events this year

[24] Simpson Papers, 4.8. Simpson Symposium, February 9–13, 2003; Tao et al., "Multiscale Modeling System"; "RAMS Description."

[25] Washington, "Warren Washington on Joanne Simpson and climate models," YouTube.

[26] Simpson, "AMS Responds to Challenges."

[27] Fleming, "Global Environmental Change and the History of Science."

hold stunning implications for our science, our profession, and for the American Meteorological Society. The Cold War has wound down. The major threat we now face is almost certain danger to our planetary environment." She mentioned the multiple stresses of an exploding population, increased pollution, drought, species extinction, stratospheric ozone depletion, and the "virtual panic about the likelihood of global warming caused by increased man-produced greenhouse gases."[28]

Focusing her remarks on the role of the AMS, Joanne pointed out the "enormous natural variability" of the air–sea–earth system on many scales of motion and modes of interaction. She explained that the Earth system poses a difficult challenge to models, "which are necessarily oversimplified," and to observations, "which are fragmentary at best," making it difficult to be confident about the rate and probable cause of global warming. She focused her presidential energies on global change science and earth science partnerships, cosponsoring a symposium on this with the American Geophysical Union and the Ecological Society of America. To emphasize the need for political action, Joanne invited Congressman George E. Brown, Jr of California to the annual meeting as a keynote speaker. Brown, who was vocal in his concerns about climate change, had been instrumental in the passage of the Civil Rights Act of 1964 and the Clean Air Act, and had helped establish the Environmental Protection Agency.

With support from NASA, Joanne welcomed thirty-nine graduate students to the meeting, initiated a program to raise fellowship money for beginning graduate students, and emphasized education of the next generation in science and math. She also expressed her hope that meteorologists learn to interact with the public more wisely than did their predecessors:

> Unfortunately, in the current media fracas about global warming, we are seeing the same excessive claims and nasty insults as we did much earlier with weather modification. It is easy, but foolish, to make exaggerated claims based on imperfect theories and incomplete evidence. It is inexcusable to try to either support or counteract these claims by personal public insults, such as "Dr. X does lousy science." This hurts all of us in our field. It makes us look ridiculous as a community. More dangerously, we lose the required credibility we must have to help decision-makers deal optimally with environmental problems, where decisions must always be made in the face of substantial uncertainty.

[28] Simpson, "AMS Responds to Challenges."

Joanne appreciated honest scientific disagreements, but regretted when it turned personal. She had seen this before with weather modification. She supported robust, even heated, but civil scientific controversies at AMS meetings, but warned her colleagues to deal cautiously with the media.[29]

Sometimes an interview reveals a previously unexplored connection. During the 1992–93 field campaign to study the interaction of the ocean and atmosphere in the western Pacific called TOGA-COARE—Tropical Oceans Global Atmosphere (TOGA) Coupled Ocean Atmosphere Response Experiment (COARE)—Peggy LeMone was stationed at Honiara, a remote airfield in the Solomon Islands with very few services. It was located near the World War II battle site of Guadalcanal. Joanne was stationed at the main logistical site in Townsville, Australia, on the other side of the Coral Sea. She and Bob had just returned from their unauthorized flights into tropical storm Oliver. At lunch one day, Peggy showed Joanne a cartoon joke booklet she had assembled about the rigors of life in the field. Joanne was not at all amused, and told Peggy, "I can't think of much that is funny about Guadalcanal. I had several friends who died there during World War II." Twenty-five years later in an on-camera interview, Peggy wondered whether the death of Joanne's friends was what inspired her to join the war effort and help train cadets in weather."[30] It was.

Virginia Gerould

Joanne never lost her childhood drive "to get somewhere and be somebody," but her mother's response to her appointments, promotions, and awards was always cool, sometimes cruel. Even when she won the coveted Rossby Research Medal in 1983, Virginia's response was, "Everyone wonders why, if you are so good, that you have not yet been elected to the National Academy." When Joanne *was* elected to the National Academy of Engineering in 1988, she received a kind word from her mother, but it turned out to be the last. Virginia became completely hostile in 1990 when Joanne was invited to send her materials to the Schlesinger Library at Radcliffe College to be part of a collection on American women's history. Joanne mentioned to her mother that the library was also interested in Virginia's pioneering role in birth control, but, Joanne recalled, "She was getting hard of hearing and persisted in

---

[29] Ibid.
[30] LeMone, "What Really Happened in Honiara…"; LeMone, "Peggy LeMone on Joanne Simpson and TOGA COARE," YouTube.

misunderstanding." When Joanne asked to send some early pictures of Virginia and her parents to the library, Virginia called angrily on the telephone and said "I don't want you to have those pictures; they are for the family." The final straw came when Joanne tried to write about her mother's life and accomplishments for the library: "She resented so much what I said that I received my last communication from her [dictated to her caretaker and] signed 'yours truly, Virginia Gerould'." Virginia had made it clear that she did not regard Joanne as a member of her family: "She refused to speak to me for the rest of her life, despite my brother's frequent pleading that she should make peace with me. In her will, I was only mentioned as the mother of my children, who received bequests. I was not concerned by a lack of [a] financial bequest, but would have felt a peaceful closure if [I had been] acknowledged as her daughter with a small piece of her delightful costume jewelry to remember her by."[31]

Virginia died at the age of 95 in 1994. She was cremated. At her instruction, no service was held. Karen joined her uncle Daniel and his son Alexander to place Virginia's ashes in the Vaughan family vault at Mount Auburn Cemetery in Cambridge, Massachusetts. Joanne tried hard to forgive the hurts and hatred she had received from Virginia throughout her lifetime and began to feel sad for her mother, who died virtually friendless with unambivalent love only for her stuffed teddy bear [Figure 1.3].[32] Joanne recalled, "I deeply regret that she died embittered toward me. I tried to reach closure as the psychiatrists recommend, namely by thinking of her as a deeply troubled woman who did the best she could under the circumstances…I have reached a far more satisfying closure from being a part of my husband's extended family and trying not to remember her in any context but that of the good things she did for me." Here is Joanne's list of her mother's good attributes, with many of the items traceable to Virginia's commitment to feminism:

1. Making it possible for me to attend the Buckingham School (except for the year in Washington, DC).
2. Acting as if it were unthinkable not to go to college.
3. Trying to have a career and pointing out obstacles to women.
4. Speaking the English language faultlessly.

[31] Simpson Papers, 1.15. Family history: Simpson's narrative: "My Mother and I," January 1998, documenting Simpson's difficult relationship with her mother and captions for family photographs from the 1930s to the 1980s.

[32] Simpson Papers, 1.12. Family history: Simpson's narrative, 1993–1994, re: mother and writing.

5. Explaining to me about birth control in a practical manner, before I reached puberty.

6. Showing by example and telling me how to avoid and deal with sexual harassment and ward off undesired passes by males.[33]

Virginia insisted on having an educated daughter. She felt strongly that education was empowerment; with knowledge comes the strength to push against and challenge the societal norms designed to keep one gender inferior to the other. Virginia was a college-educated woman who sought to teach future generations of women, including her daughter, how to succeed in the male-dominated workplace while maintaining control of their bodies and their fortunes. Joanne appreciated her mother's feminist activism and adherence to liberal ideals, which she credited with forming the foundation of her own personal autonomy and ability. Joanne consistently tried to ignore any kind of misinformed name-calling or belittlement. Instead, she focused on her work. She compared her approach to that of her stepdaughter, Peg, whom she described as "an outspoken, uncompromising, fighting feminist" in her mid-60s, but without a job. "All her life Peg marched in all the parades, corrected her father or friends when they said something like 'hire a man' when they meant 'hire a person', and made a big fuss about naming hurricanes after women. She led a class action discrimination suit against the Associated Press. They won the suit, each woman got about $10,000, but they all were forever blacklisted in journalism. She hasn't had a job since, and has to make what she can freelancing."[34] Joanne was deeply passionate, however, about some big causes. In the 1960s she marched in the streets in support of reproductive choice and in protest of the war in Vietnam. At age 80 she was prepared to march in the massive demonstrations against the Iraq War, but her poor health prohibited it.

Return to the Realm of MONEX

In January and February 2001, to celebrate their thirtieth-sixth anniversary, Joanne and Bob took a seven-week cruise through Southeast Asia, India, the Middle East, Egypt, and the eastern Mediterranean.[35] Their

[33] Simpson Papers, 1.15. Family history: Simpson's narrative: "My Mother and I," January 1998, documenting Simpson's difficult relationship with her mother and captions for family photographs from the 1930s to 1980s.

[34] Joanne Simpson to Peggy LeMone, June 9, 2003, e-mail, author's personal collection.

[35] Simpson Papers, 2.1. Travel diary, scrapbook, and itinerary, January–February 2001.

vacation covered much of the area that had been dedicated to the 1978–79 MONEX program to study the effects of monsoons at the inter-faces of tropical land areas and oceans. They flew to Bangkok, Thailand, to meet their ship, and then stopped in Vietnam for three days. Here Joanne reflected on her son Steven's tour of duty twenty-five years earlier. Not one to join crowds, she had marched against the war—once. They crawled through the tunnels in Cử Chi, near Ho Chi Minh City, that had housed up to 16,000 Viet Cong, calling it "an impressive lesson in guerrilla warfare." This reinforced her original view that the United States never should have interfered in the first place, nor should it have backed a corrupt dictatorship against the will of the majority. Our government claimed that we had to get rid of "communism" or all the "dominoes" would fall, and then they lied to us about the "light at the end of the tunnel" and the numbers of dead on each side.

Bob and Joanne's overwhelming first impression of life in Ho Chi Minh City was the two million motorbikes that drive on any side of the road they please, resulting in ten deaths per day on average. The ship made stops in Malaysia in Kwantan, Singapore (where they noted that the Singapore Slings did not have enough gin), Malacca, Kuala Lumpur, and Penang Island, sites of their winter MONEX field program over two decades earlier. Next stops were Myanmar and South Andaman Island, the site of a former British penal colony. Their scheduled port call in Sri Lanka was cancelled due to political unrest, and instead they docked at Cochin, located on a large tidal estuary on the southwest coast of India.[36]

Joanne suffered a serious misfortune in Mangalore, India, while on a bus excursion. Her hip replacement separated, and she could not pop it back in, which caused her excruciating pain. One of their fellow travel-ers, a recently retired orthopaedic surgeon, also tried, but could not reassemble the hip joint. Treatment required a long, bumpy ambulance ride to a local hospital where, after X-rays and under anesthesia, the doctors repaired Joanne's hip and supported it with a huge metal brace. The next stop was Mumbai, with its beautiful Gate to India illuminated in the pink light of the setting sun, followed, after a passage through the Arabian Sea, by Dubai in the United Arab Emirates and Oman, a place Joanne had once visited to consult on water issues. The ship tra-versed the Gulf of Aden and the Red Sea, skipping its port of call in

[36] Ibid.

Yemen due to political strife, and docked four days later in Safaga, Egypt, which resembled an armed encampment.

Joanne really enjoyed their excursion to Mount Sinai, where they visited St. Catherine's Monastery, which seemed to them to be a tourist trap. They had a more authentic experience drinking tea and eating flat-breads in a nearby Bedouin encampment. They traversed the Suez Canal and arrived in Port Said the following day, crossing the Mediterranean Sea to Crete. "It was wonderful to be in Europe!—Not a policeman nor military uniform in sight."[37] They stopped in Kusadasi, Turkey, where Joanne fell on some stone steps and hurt herself, but not seriously. The voyage ended in Piraeus, Greece. They had enjoyed sailing for seven weeks throughout the continental tropics, enjoying its airs and discussing its monsoonal flows. They had seen many sights and experienced a variety of cultures. Joanne had two accidents, one of which was major. She reflected on the cultures and religions they had encountered, deciding she liked Buddhism the best, because "they don't kill people who don't agree with them."[38] In her opinion, the Middle East was dominated by despair, anger, and distrust, all stoked by the rapid geographical expansion of Islam and the violent behavior it seems to breed. She feared for the future.

Joanne also had reason to fear for her husband's future. That summer, as Bob, aged 89, was preparing his *Hurricane!* book, he became seriously ill and required heart surgery. He was suffering from arteriosclerosis and required two additional stents to accompany the two he had received the previous year. He was very weak after the surgery, and during his long hospital stay, resigned himself to the probability that he was near the end of his life. The doctors told Joanne and his daughters that he might not make it until Christmas. It was a tough time for all. Since Joanne was eleven years Bob's junior, she expected to outlive him but did not want to go on without him. To prepare for what seemed to be the inevitable, she joined the Hemlock Society—a right-to-die organization that supported legalizing physician-assisted suicide.[39] But Bob made a full recovery, and with the assistance of Joanne and two

---

[37] Ibid.
[38] Ibid.
[39] Malkus-Benjamin, "Video Interview," by Lipshultz.

associate editors, Richard Anthes and Mike Garstang, the book was ready in 2002.

In 2002 Joanne received welcome news that she had won the IMO Prize of the World Meteorological Organization (WMO) "for outstanding work on the science of meteorology." It came with a gold medal and a cash award of 10,000 Swiss francs. She called this "an apex event" in her life. She received the award, the first ever given to a woman, in the Great Hall of the National Academy of Science in Washington, DC, with more than a hundred colleagues, friends, and family attending the reception and buffet dinner. She felt "overwhelmed" to be joining the company of previous award winners such as Rossby, Riehl, satellite meteorologist Verner Suomi, and the founder of chaos theory Ed Lorenz.

Joanne was apprehensive about two aspects of the ceremony: standing for more than twenty minutes due to the pain of her recent complete hip joint replacement surgery, and forgetting to acknowledge someone during her acceptance speech. The actual affair, however, went smoothly. Bob pinned a corsage of white roses on her lapel. Her adrenaline flow kept her happily standing and moving around the gathering for more than an hour during the reception, talking animatedly to everyone and basking in all the attention from her long- cherished colleagues. She recalled that the high point was meeting Rita Colwell, a famous microbiologist who headed the National Science Foundation: "She had come there just to meet me! I told her she was my heroine and she said I was hers! She asked me to send some of my reprints to her, while we had a meaningful conversation about how we kept our science going when our kids were little."[40] The after-dinner program included remarks by Patrick Obasi, Secretary General of the WMO, who spoke of Joanne's research accomplishments, and WMO President John Zillman, who made the actual prize presentation. Franco Einaudi, Joanne's colleague at NASA and Director of the Earth Sciences Division, told a story about how a commercial airplane taking Joanne to an AMS meeting in Boston had to reduce its altitude because she had recently had eye surgery, which would have been undone if the cabin pressure fell

[40] Simpson Papers, 2.7. Awards: International Meteorological Organization Prize, 2002, includes press release, Simpson's narrative, acceptance speech, event program.

too low. His point was that Joanne was so hypercompulsive about her obligations that she had endangered her health to keep them. Her acceptance speech received a standing ovation, her first. At the end of the ceremony she was handed a huge bouquet of long-stem red roses, a gift from Bob, which she received in the manner of an opera star, bowing to the audience and telling them she regretted she could not carry a tune. She collapsed the next day with a ferocious migraine headache, but after a day of rest the glow of achievement returned. She received a number of formal congratulatory notes from people whose lives she had touched. Joyce Annese, Director of Executive Programs at AMS, sent a personal note. Joyce was the staff member in charge of all meeting arrangements when Joanne had held positions of authority in the Society. Joanne recalled, "I think that she, as well as other members of the AMS staff, got fond of me because I interacted with them as humans, tried to make it visible that their contributions were important, and thanked them personally for what they did to help me and the Society. Some of the other Presidents either totally ignored the staff or treated them as lowly clerks or servants, which they were definitely not. Since many women resent women whom they perceive as in a 'higher' status, I am very grateful for Joyce's affection, which I can tell is real."[41] Joanne used the IMO prize money to purchase several hundred copies of Bob's AGU *Hurricane!* book for distribution to colleges, universities, and weather services around the Caribbean and other less wealthy nations affected by hurricanes.[42]

Joanne found personal happiness and received numerous honors late in life. NASA was active in recruiting and promoting women scientists, providing a unique work atmosphere that did not isolate women from their male colleagues, and interdisciplinary cooperation was encouraged. Abundant new data was streaming in from TRMM and new insights from the cloud models. Joanne's marriage to Bob was a happy one, and her personal contentment paralleled her professional success. She officially retired from NASA in 2004, but came into the office as long as she was physically able (Figure 8.5).

[41] Ibid.
[42] *Hurricane!*

Figure 8.5 Joanne in her office at Goddard, 1997. Simpson Papers, 455, Photos. PD-19. Credit: T. Nohe, NASA/NSSDC.

NASA Exit Interview

Joanne completed her exit interview at NASA on September 3, 2004, twenty-five years and one day after she arrived there. The interview, a standard practice, was aimed at building institutional memory and learning from the experience and insights of the retiree. Here are the eight questions and Joanne's responses:[43]

Which projects do you consider your favorites or the most successful? Why?

Favorite and most successful: 1) TRMM satellite and 2) numerical cloud modeling. Why? TRMM has been a vital science contribution, an

---

[43] Simpson Papers, 3.6. Interview transcript re: Simpson's exit interview from NASA, September 3, 2004.

important help to forecasters, and a near perfect NASA Project; launched on schedule, within budget, and successful data taking. Satellite has now been [in orbit] more than twice the three years planned. All goals reached or exceeded. Cloud modeling has gone from infancy to applicable maturity during my 25 years at NASA. Wonderful colleagues. New basic physics insights, and [new] TRMM algorithms have come from our cloud models, thus the modeling effort has been necessary for TRMM success. The most advanced cloud-resolving models are necessary for global weather and climate predictions, thus of value to the taxpayer.

What are the five key lessons that you have acquired over your NASA experience that you would like to share with others?

1. Always think ahead, planning and conducting new research for the next proposal before the date of Headquarters Announcements.
2. Keep in touch with Headquarters Program Managers and try to have inputs they will incorporate into the direction of their programs.
3. Make your work known to science colleagues, get as many involved in the work as possible with the always limited resources. Later, their influence counts with Headquarters and continued funding for work I know is feasible and important.
4. Delegate to others as much as possible, especially paperwork. Thank people who help and show appreciation to them. Give support people, who deserve it, praise to their supervisors and small gifts from trips and other tokens of gratitude. People can rarely be persuaded by iron fists or getting tough with them. Express clever flattery, implicitly if possible. Learn to say "no" tactfully.
5. Listen and don't talk too much. I am still working on this one.

Which projects, if any, resulted in some type of failure? Why?

No total failures. All scientists have to become accustomed to the fact that some of their ideas will be wrong and will need modifications as they develop. Some projects have a dead end at some point. My work on waterspouts was less than great and did not cause as much discussion as I had wished. No one in NASA was interested in following up on it with me. It ended after two journal publications, which my Australian colleagues participated in with me. The work was a big part of the reason I was invited to be a Visiting Professor at Monash University in Australia, which proved useful to NASA and to TRMM.

*What kinds of questions did you ask or were asked by your
colleagues of you on a typical working day?*

There is no one typical working day. Those days when I can really think
about science for a useful period of time, sometimes with a colleague,
are the best and were not as frequent as I wished after 1986, when I took
over the lead of TRMM. During TRMM, presentations and project
meetings and subgroup meetings took nearly full time. There were
three major hassles where high-level people tried to kill TRMM.
Fending them off required huge efforts and the marshaling of all outside supporters. All told, these struggles and the numerous presentations involved took at least one year of full time work on my part. The
first two spread out between about 1987–1991 when TRMM finally
became a project. The last was an attempt to take away half of the data
system budget in the last months before launch in 1997. Questions
I asked and was asked during TRMM related to everything from cloud
modeling to details of the Project budgets and sub-budgets, the main
purposes of TRMM, and new ideas that came up about instruments,
data systems, planetary atmospheres, climate variability, how TRMM
would help understand global warming, etc.

What role did you play on your team?

I served as leader or co-leader of the science parts of the major projects
or teams I have been involved with at NASA.

*Would you be interested in mentoring some junior
Goddard colleagues upon your retirement?*

Mentoring is one of the favorite roles that I have played for more than
50 years. I will be pleased to continue on a more limited scale.

*Would you mind being contacted by our Goddard knowledge
management team sometime in the near future for follow up?*

Further contact on these subjects will be welcome.

*Would you be interested in being videotaped as part
of Goddard's knowledge preservation project?*

[No response][44]

[44] Ibid.

Civil Discourse on Climate Change

In 2008 Joanne posted a weblog "as a private citizen" to provide her hard-won perspectives on contentious scientific issues in the public realm. She again deplored the lack of civility surrounding climate controversies and promoted the use of the TRMM data set for resolving climate issues.

Since I am no longer affiliated with any organization, nor receive any funding, I can speak quite frankly. For more than a decade now, "global warming" has become the primary interface between our science and society. A large group of earth scientists, voiced in an IPCC statement, have reached what they claim is a consensus of nearly all atmospheric scientists that man-released greenhouse gases are causing increasing harm to our planet. They predict that most icepacks including those in the Polar Regions, also sea ice, will continue melting with disastrous ecological consequences, including coastal flooding. There is no doubt that atmospheric greenhouse gases are rising rapidly and little doubt that some warming and bad ecological events are occurring. However, the main basis of the claim that man's release of greenhouse gases is the cause of the warming is based almost entirely upon climate models. We all know the frailty of models concerning the air–surface system. We only need to watch the weather forecasts. However, a vocal minority of scientists so mistrusts the models and the complex fragmentary data, that some claim that global warming is a hoax. They have made public statements accusing other scientists of deliberate fraud in aid of their research funding. Both sides are now hurling personal epithets at each other, a very bad development in Earth sciences. The claim that hurricanes are being modified by the impacts of rising greenhouse gases is the most inflammatory frontline of this battle and the aspect that journalists enjoy the most. The situation is so bad that the front page of the *Wall Street Journal* printed an article in which one distinguished scientist said another distinguished scientist has a fossilized brain. He, in turn, refers to his critics as "the Gang of Five."

Few of these people seem to have any skeptical self-criticism left, although virtually all of the claims are derived from either flawed data sets or imperfect models or both. The term "global warming" itself is very vague. Where and what scales of response are measurable? One distinguished scientist has shown that many aspects of climate change are regional, some of the most harmful caused by changes in human land use. No one seems to have properly factored in population growth and land use, particularly in tropical and coastal areas.

What should we as a nation do? Decisions have to be made on incomplete information. In this case, we must act on the recommendations of Gore and the IPCC because if we do not reduce emissions of greenhouse gases and the climate models are right, the planet as we know it will, in this century, become unsustainable. But as a scientist I remain skeptical. I decided to keep quiet in this controversy until I had a positive contribution to make. That point is to be celebrated in the TRMM 10-year anniversary in a conference in February 2008. With a 10-year record [of] TRMM, users of the data can begin to look for and test for trends … in global tropical rain on several scales, including regional.

These patterns can be compared over the past ten years with the patterns predicted ten years ago by the climate models. An example is the Walker Circulation, normally with descent of air over the eastern Pacific Ocean and ascent of air over the western Pacific. When this cell weakens, perhaps breaking over the middle Pacific, we have an El Niño. The modelers say that higher greenhouse warming produces recognizable changes in the Walker Circulation. What better data is there to test such model results than the tropical rain products from TRMM? While the TRMM data set provides no panacea on the volatile hurricane front, useful information for the several ocean basins relating the rainfall to claimed and observed storm structure can be made if dedicated work is committed. I would be most interested to find out how the distribution of hot towers relates to storm intensity and rain production. Examining the data already posted on the TRMM website shows that such projects are tractable. The major lack for TRMM data use in testing climate theories is latitude limitation. Global warming impacts appear much more severe in polar latitudes than in tropical regions. The best news is that the Global Precipitation Mission (GPM) is on schedule for a 2013 launch. In conclusion, I can just pray that GPM scientists and engineers are as smart and as lucky as we TRMM participants have been.[45]

Joanne had inserted a voice of reason into an overheated controversy.

Remembering Joanne

Jeff Halverson first met Joanne flying at 40,000 feet over the Coral Sea during TOGA-COARE and was immediately impressed by her intensity. He worked with her as a postdoc for seven years. He said that he stayed there for such a long time "*just because she was there,* and we were working on exciting things … We were flying into hurricanes; we were collecting the data; we would come back and analyze the data." They would spread out, organize, and analyze the maps, charts, and images on the floor of her "huge, huge office [Goddard Building 33,

[45] Simpson, "TRMM (Tropical Rainfall Measuring Mission) Data Set Potential."

Room C407]...Just get down on the floor with your mentor and look
at stuff, and we didn't care who was walking by...We could lose
ourselves for hours talking nothing but science and looking at data.
She had the most amazing amount of enthusiasm and motivation, and
it was infectious! I caught it; I caught the bug, and then I would provide
feedback to her because she saw my excitement and it kept her going.
And so that's how we published a lot of our papers."[46]

After the failure of her hip replacement surgery in 2001, Joanne spent
the last few years at work zooming around Goddard in a motorized electric scooter sporting an orange pennant and inspiring her younger colleagues with her seemingly endless ideas. She seemed to be always in a
hurry, and there are still tire marks visible on the floor tiles and walls
where she had mistimed her turns. At this time, Goddard installed con-
vex mirrors on the hallway corners to reduce "traffic" accidents. She once
told Peggy LeMone, "My greatest wish would to be like Grady Norton,
who died of a heart attack while forecasting a hurricane, or like my early
hero, Rossby, who keeled over and died in the middle of giving a seminar.
I don't like the idea of when I won't be a meteorologist anymore. It's just
inconceivable to me."[47] When asked in 2010 if she would do it all over
again, if given the chance, she replied, "I'd do just the same thing."[48]

Joanne Simpson was terminally ill, but, as described by her daughter
Karen, "She was fighting to be with Bob in the end rather than wanting to
go away."[49] The end came, the result of multiple organ failure, on March 4,
2010. She was 86.[50] The following day, a number of her colleagues shared
their reminiscences. Richard Anthes, one of her early collaborators in
Florida, and president of the University Corporation for Atmospheric
Science (UCAR) at the time, remembered her remarkable "energy, enthusiasm, intelligence, and support for the people she worked with," and how
much she enjoyed informal social interactions after hours with her many
friends and colleagues. He referred to her as "a role model and mentor for
all young and not-so-young scientists and humanists," adding that she was

[46] Halverson, "Jeff Halverson on Joanne Simpson and working together," YouTube.
[47] Simpson, "Interview," by LeMone.
[48] "First Lady of Tropical Meteorology."
[49] Malkus-Benjamin, "Video Interview," by Lipshultz.
[50] Sullivan, "Joanne Malkus Simpson." The obituary erroneously referred to
Simpson as the "first female meteorologist to earn a doctorate," an error repeated by
the *Boston Globe*, the *Wall Street Journal*, the *Bulletin of the American Meteorological Society*, the
journals *Nature* and *Weather*, *Memorial Tributes of the National Academy of Engineering*, and even,
in her old age, by Joanne herself.

"generous, kind, sparkling, caring, inspirational, vivacious, determined, unpretentious, forgiving, and loving."[51] Peggy LeMone—Joanne's close ally and associate on women's issues and the 2010 president of AMS—called her "an inspiration to women in meteorology for over a half century." Joanne had set the example that made it possible for Peggy and other women to imagine themselves as meteorologists.[52] Greg Holland, one of Joanne's collaborators in the TOGA-COARE program, referred to her as "one of the giants of tropical meteorology, a remarkable advocate for women in science, a great mentor to many young scientists, and a dear friend."[53] Roger Pielke, a long-time collaborator wrote: "The richness of Joanne Simpson's research accomplishments are best appreciated by tracking our current knowledge of the atmosphere to where these concepts were first discussed in the peer-reviewed literature. Her breadth of contribution is impressive and ranges from the cumulus cloud to global scale. Early in her career, she recognized the critical role of cumulus clouds in the earth's atmosphere... When one uncovers the origin of many of our most basic concepts in atmospheric science, it is quite impressive how much of this knowledge is founded in her original work!"[54]

Joanne and Bob Simpson joined their lives, their interests, and their families in a marriage that lasted four-and-a-half decades. Joanne remarked in 2010: "People thought when Bob and I got married that we'd be in competition with each other and the marriage wouldn't last more than two years. We found out later they took bets on it, and now we're in our 45th year of our marriage... and I don't think we've *ever* had any competition between the two of us. It's one of us putting in an idea and another one developing it and going back and forth."[55] Bob felt the same way: "My association with Joanne Malkus, who became Joanne Simpson, has meant more to me both professionally and personally than any other factor in my life. There were many things that Joanne did really well that I couldn't touch, and I believe some things that were easy for me and more difficult for her I was able to help with. In any event our lives have been uniquely buoyed by our decision to meld our two careers through marriage."[56]

[51]   Anthes, "Remembering Joanne Simpson."
[52]   LeMone, "Remembering Joanne Simpson."
[53]   Holland, "Remembering Joanne Simpson."
[54]   Pielke, "Joanne Simpson—An Ideal Model of Mentorship."
[55]   "First Lady of Tropical Meteorology."
[56]   R. H. Simpson, "Interview," by Zipser.

9

Breaking Through

> If one advances confidently in the direction of HER dreams, and
> endeavors to live the life which SHE has imagined, SHE will meet
> with a success unexpected in common hours.
>
> ADAPTED FROM THOREAU

Joanne Simpson was not the first woman meteorologist, nor was she the
first to earn a PhD in the field, but she was the very best tropical meteorologist of her generation, the first to achieve legendary status. She was a
highly accomplished mentor and an initiator of group collaborations,
from her fundamental work on hot towers with Herb Riehl to the
highly successful Tropical Rainfall Measuring Mission satellite, and she
is credited with many fundamental scientific and professional accomplishments during her long career. These include leadership of large airborne observing projects, computerized cloud models, understanding
trade wind convergence, the hot tower hypothesis of cumulus convection, experimental seeding of hurricanes and clouds, and supervision of
TRMM. Her stature and influence resulted in a marked increase in the
number of female meteorologists over the course of her life. She was not
just a pioneer female scientist, she was a pioneer scientist, period. She
was a determined professional woman trying to break into what had
been an all-male field and open doors for others, with all that that
entailed. While this required immense effort and personal sacrifice, it
came with great rewards. It made her a stand out, a highly visible central
figure in the new field of tropical meteorology she had done so much to
create and nurture. Researchers continue to cite Joanne's work. A leading textbook on tropical meteorology published in 2013 features her hot
tower hypothesis, her computerized cloud model, her work on TRMM,
and refers to her as "a pioneer in bringing out the importance of tropical
clouds as a distinctly different entity from [those in] middle latitudes."[1]

---

[1] Krishnamurti et al., *Tropical Meteorology*.

First Woman: Joanne Simpson and the Tropical Atmosphere. James Rodger Fleming,
Oxford University Press (2020). © James Rodger Fleming.
DOI: 10.1093/oso/9780198862734.001.0001

Many men (and one woman, Anne Louise Beck) established the Bergen School of forecasting in the 1920s; many men (and one woman, Inger Bruun) served as state meteorologists in the 1940s; but Joanne (and one man, Herbert Riehl) pioneered the field of of tropical meteorology. In 1953, Rossby's comment to Joanne that she was "sticking out like sore thumb" was insensitive. But there was an element of truth in it: she did stick out, or rather stand out, not as a sore thumb, but as a unique success. The reality of her life was much more complex than anyone understood. While her accomplishments are beyond question, her personal life was tumultuous, and the professional barriers she faced were daunting. She worked at the cutting edge of a new field, published under three married names, raised three children, and successfully confronted her health problems and personal insecurities, while reaching for the top and opening doors for others.

In 1990, as Joanne was preparing her archival donations to Radcliffe, she reflected on the struggles of the past. She concluded that the obstacles placed before her because of her sex were actually immense benefits in disguise. She realized that being rejected in the mid-1940s for mainstream jobs in meteorology and subsequently ending up on the physics faculty at the Illinois Institute of Technology was one of the best things that ever happened to her. The scope and intensity of her teaching responsibilities broadened her scientific range and helped develop her pedagogical and communications skills. She also thought it was beneficial to have been an unknown junior scientist at Woods Hole, somewhat out of the mainstream until 1960. Because of this, she had time to think and develop ideas without interruption. This led to her best work.

After 1960, as her career began to take off, outside demands on her time increased: committee work, project and program leadership, supervision of young people, and increasing expectations to serve as a role model for women (Figure 9.1). She considered the latter responsibility a duty, a privilege, and a very heavy burden. Her mantra was, "If I mess up, every other woman will suffer for it." It was only after she had won the Rossby Award, gained election to the National Academy of Engineering, served as President of AMS, and won the IMO Prize, that she felt "almost safe," with some leeway to go "off the rails" without damaging the careers of her female successors. Even into the 1990s, however, she was subjected to infuriating statements from male colleagues such as "You are the exception that proves the rule." She had made major contributions to the field and was ready to retire in

Figure 9.1 Joanne Simpson, *c.*2004. Simpson Papers, 455, Photos. PD-15. Original in color.

peace from being a role model. Other women would win major awards in meteorology, but that would come later—quite a bit later (Table 9.1).[2]

In 1974 only 4% of the professional members of AMS were women, but by 1999 that figure was 11%, and by 2017 it was 22%, with 43% female student members.[3] Now the field is becoming more diverse, enrolling equal numbers of women and men, with one in every three PhDs in meteorology earned by women.

Most people cite the relationship with one or both parents as an important or defining influence on their emotional life, their way of looking at the world, and their earliest feelings of love and self-worth. Joanne lacked this stability and instead, focused on academic pursuits, where she excelled. Her reward structure derived from friends and

---

[2] J. Simpson, Papers, 9.2. Narrative by Simpson, January 1994, re: description of undergraduate and graduate college career.

[3] Hartten and LeMone, "Evolution and Current State"; Behl et al., "Diversity at AMS."

Table 9.1 Joanne Simpson's major awards and honors in meteorology, name and date of the second female awardee, and the number of years intervening.

Award	Awardee	Date	Years to next female awardee
Rossby Medal	Joanne Simpson	1983	
	Susan Solomon	2000	17
National Acad.	Joanne Simpson	1988	
Engineering	Eugenia Kalnay	1996	8
AMS President	Joanne Simpson	1989	
	Susan Avery	2004	15
IMO Prize	Joanne Simpson	2002	
	Eugenia Kalnay	2009	7

teachers at school, "who provided affection, guidance, and extremely high standards of scholarship, and who inspired curiosity concerning all intellectual pursuits."[4] Much later in life, she was incorporated into Bob Simpson's extended family where she experienced their warm and devoted feelings toward each other.[5] Joanne struggled to find love and acceptance, waged a constant battle with bouts of depression and frequent crippling migraine headaches, and confronted institutional obstacles like nepotism rules and structural sexism. But she did not give up. She helped those coming after her through her example, advice, and actions.[6] As she gained job security and seniority at NASA, Joanne became more open about her struggles with mental health.

> My greatest lifelong battle has NOT been being a woman. It has been with moderate to severe depression, which runs in my family. Until I reached 70 I never dared to confide this to anyone but my husband, then I began to discuss it more freely with my children . . . I also spoke about it with one or two colleagues who appeared to be going through the same ordeal, to see if I could help them, as apparently, I have in one case at least. Now there is no longer need to be afraid of the stigma, since my

---

[4] J. Simpson, "Meteorologist."

[5] Simpson Papers, 1.15. Family history: Simpson's narrative: "My Mother and I," January 1998, documenting Simpson's difficult relationship with her mother and captions for family photographs from the 1930s to the 1980s.

[6] LeMone, "Remembering Joanne Simpson."

54-year career in meteorology has been recognized as outstanding, and I am no longer in a position where I might be looking for another job.[7]

Since Joanne wrote for and regularly read *Scientific American*, she probably saw a 1995 article on manic-depressive illness and creativity that featured eighteen writers and artists, only four of them women, but none of them scientists. Joanne reflected on the article's points regarding the inherited nature of the disease, its extreme rarity in the general population, and its frequent occurrence among writers and artists.

> The stigma that attaches to even a few visits to a psychiatrist, added to being a woman and being divorced, were more than I thought my career opportunities could stand. Today in the world of science, this situation seems to be improving. Also, recently a correlation between manic-depressive personality syndrome and creativity has been demonstrated in authors and in artists. I believe there is frequently a similar association in creative science.[8]

A Celebration of Research

Given all the pain, disappointment, and mental anguish she experienced, Joanne could have given up on life. Instead, in her seventy-sixth year, in 1999, NASA celebrated her career in research, management, and mentoring in a two-day symposium marking the fiftieth anniversary of her PhD.[9] The speakers—some of them Joanne's students, and all of them Joanne's colleagues—provided historical and personal perspectives as they reviewed the progress and current state of knowledge of clouds and the tropical atmosphere. The papers appeared in a peer-reviewed monograph, and four years later in a book.[10] She wanted to make it clear that scientists alone cannot design, build, and successfully launch a satellite. There had to be very close teamwork.

Goddard Center Director, Alphonso Diaz, opened the symposium by acknowledging Joanne's many accomplishments, including her role as TRMM project scientist, her authorship of over 190 refereed

---

[7] Simpson Papers, 1.14. Family history: Simpson's narrative re: difficult childhood, lifelong depression, detailed photograph captions with commentary, January 1996.

[8] Jamison, "Manic-Depressive Illness and Creativity"; Simpson Papers, 1.14. Family history.

[9] Tao et al., "Summary of a Symposium."

[10] Tao and Adler, eds., *Cloud Systems, Hurricanes, and the Tropical Rainfall Measuring Mission (TRMM).*

publications, and her ability to bridge the fields of science and technology. He praised her as a role model for scientists and for her dedication to service, both nationally and internationally. Next, Vincent Salomonson, director of Earth Sciences at Goddard, pointed out her important role in securing TRMM funding early in the program, noting with pride that she consistently outperformed most other employees in her annual performance appraisals, and that the most powerful computer at Goddard was named in honor of her. The keynote speaker was Ghassem Asrar, Associate Administrator for Earth Science at NASA Headquarters, who pointed out that her role in TRMM was just one example of her approach to life and science: do everything fully, keep the big picture in mind, and let the "smart" people, as she called them, handle the intermediate steps. He noted that although Joanne was the first woman to obtain a PhD in meteorology in the US, she never used that distinction for herself, and always tried to ensure that women got their fair shot in science.[11]

Radar expert and Joanne's former boss, David Atlas, spoke of her "phenomenal enthusiasm for all of meteorology" and called her a "mythical figure." Her cloud-seeding efforts in Florida helped her provide good advice to the NHRE, where hail suppression research was the focus. At NASA he expressed his admiration for her "inspiring enthusiasm, bottomless well of good ideas, ability to attract bright young people who remained loyal to her, and one absolutely necessary quality in a big bureaucracy—the ability to stand up to program managers at NASA headquarters better than anyone else."[12]

Several distinguished women scientists spoke next. All of them were Joanne's friends, all had been inspired by her, and three followed her into the National Academy of Engineering. Eugenia Kalnay, environmental modeler and first woman to earn a PhD in meteorology from MIT, expressed her gratitude that she was able to follow in Joanne's footsteps. Joanne responded in good humor, stating that although she had been "the role model for women in meteorology," she was resigning that position and passing the mantle on to some of the women scientists in the audience.

Kristina Katsaros, director of NOAA's Atlantic Oceanographic and Meteorological Laboratory, referred to Joanne as the "number one woman of air–sea interaction research." It was a mutual admiration society. Joanne wrote, "Kristina is probably the most successful and

[11]   Simpson Papers, 4.8. Simpson Symposium, February 9–13, 2003.
[12]   Tao et al., "Summary of a Symposium."

productive of the 'second generation' of women meteorologists."
Kristina earned a PhD in atmospheric science at the University of
Washington in 1969. When she wrote a "fan letter" to Joanne after reading
her paper on sea–air interactions, Joanne responded with a copy of an
article on her struggles as a divorced woman scientist.[13] Kristina
accepted this as wise advice for her own career and marriage, and
Joanne became her mentor in a field with very few professional women.
Joanne and Kristina worked together in Miami, and Joanne nominated
her for election to the National Academy of Engineering. Kristina
remarked, "She always had great ideas about inclusiveness and important
kind and good things to do for other people. She was a mentor not just
to me, but to many people." At Joanne's 2003 retirement party at
Goddard, Kristina presented Joanne with a cartoon she and her husband
had prepared for the occasion (Figure 9.2). She particularly admired the
way that Joanne fought for funding for her satellite, TRMM.[14]

Figure 9.2 "Joanne" by Kristina and Michael Katsaros, 2003: the optimistic
woman knocking on the door, the intrepid woman knocking down the wall,
and the wise woman flying over the wall. Image provided by Kristina Katsaros.
(Katsaros, "Kristina Katsaros on Joanne Simpson Cartoon.")

[13] Simpson, "On Some Aspects of Sea–Air Interaction"; J. Simpson, "Meteorologist."
[14] Katsaros, "Kristina Katsaros on Joanne Simpson as her Guardian Angel," YouTube.

Peggy LeMone, Joanne's close friend and colleague from NCAR, spoke about women in meteorology, comparing the results of the survey that she and Simpson had conducted in the early 1970s to more recent data, which showed an encouraging increase in the number of women in the atmospheric sciences and a reduction in the problems they encountered in sustaining their careers—a trend she attributed to Joanne's strong mentorship. She closed by citing the potential additional benefit of women in science because women can bring a different set of experiences and hence a different outlook. Peggy also shared a retirement song, sung in Joanne's honor at her retirement party.[15]

Joanne was an equal opportunity mentor. Mike Garstang's association with Joanne dated from the 1950s during her Woods Hole days. They subsequently became faculty colleagues at Virginia after Garstang nominated her for a job there. Bill Woodley, one of Joanne's undergraduate students at UCLA, worked with her in Florida, and remained a close associate over the years. Gregarious and fun-loving, his sense of humor enlivened the field campaigns and the parties. Rick Anthes met Joanne in the 1960s when he worked as a student trainee at the National Hurricane Research Laboratory in Miami. He remembers attending seminars there and seeing her sitting in the front row. "Bob and Joanne would sit together, arms around each other, holding hands, snuggling, and kissing, like love birds . . . They were probably in their fifties at the time and were carrying on like teenagers in front of everybody, so they became the butt of a lot of jokes among us young people." Rick's story illustrates how deeply in love Bob and Joanne were. He adored her, and she adored him.[16]

Joanne referred to Roger Pielke as "one of my best mentees . . . a pioneer in mesoscale modeling of cloud systems and their environment." Roger published an article on her "ideal model of mentorship." As a PhD student at Penn State University in 1971, he joined EML, where Joanne provided the environment, resources, and intellectual stimulation for him and other young scientists to flourish. Although primarily a modeler, he flew with her in the FACE project and on hurricane flights before moving with her to the University of Virginia as an assistant professor. There, he adopted her philosophy of serving as a facilitator for student research, rather than as a manager. This approach emphasized exposing students to the latest ideas and research practices through

[15] LeMone, P. "Peggy LeMone on Joanne Simpson and Women in Meteorology," YouTube; "Peggy LeMone on Joanne Simpson NASA Retirement Song," YouTube.

[16] Anthes, "Rick Anthes on Joanne Simpson and Bob as Lovebirds," YouTube.

collegial interactions. Pielke recalled later, "Joanne gave students enough rope to be creative, but not enough to hang themselves."[17]

Bill Cotton earned his PhD at Penn State in 1970, and faced with several options, chose to work with Joanne at NOAA, which Cotton said "was a really good move...Probably you've heard this before, but Joanne was one heck of a mentor. She was a role model for [my career working at Colorado State University with over forty-five PhD students]...She would go to bat for you, for her people."[18] Cotton developed a model that linked small-scale convection with mesoscale processes that force the convection. He enjoyed healthy scientific disagreements with Joanne and encouraged cloud modeler Roger Pielke to join the NOAA team. The two men wrote a book that compared weather modification studies with the current overselling of the ability to predict future climate. Their book emphasized, as had Joanne, that hot towers deserve more study. Hot towers cover a very small percentage of the Earth's surface but have a very large effect on climate, and are critical to our understanding of human-caused climate sensitivity.[19]

In 1987, Steven Rutledge, a young assistant professor at Oregon State University, responded to Joanne's call to help on a project in Darwin, Australia, aimed at calibrating the TRMM radar far in advance of its launch. Her invitation got him in on the ground floor of the project and "opened up field work in the tropics" for him. "I know I'm not the only young scientist that Joanne got involved in things and helped jumpstart their career. And I remember I sent her a letter, shortly before she passed away, thanking her for the opportunity, and she sent me a very kind note back that I still have today. I always will treasure that. She was a wonderful person, an incredible scientist, an inspiration to us all, and I'm grateful for the opportunities she gave me."[20]

Joanne praised the work of Chris Kummerow, the "unsung hero" of TRMM and its "best treasure." She recruited Chris out of the University of Wisconsin to develop the TRMM data system. She wrote, "If I believed in Divine Providence, I would thank it for getting young Chris

[17] Pielke, "Joanne Simpson: An Ideal Model of Mentorship"; Pielke, telephone interview with Fleming, July 12, 2018.

[18] Cotton, "Bill Cotton on Joanne Simpson, Why He Chose His Graduate School," YouTube.

[19] Cotton and Pielke, *Human Impacts on Weather and Climate*.

[20] Rutledge, "Steven Rutledge on Joanne Simpson," YouTube; email to the author, Oct. 10, 2017.

to Goddard."[21] Kummerow, who took over as TRMM Chief Scientist after the satellite was launched, said that he will never forget how Joanne both protected him and then corrected him after he gave his first talk about the project and then received severe criticism from senior NASA officials:

> I was just coming up the ranks. It was the first time she asked me to give a talk at a conference for the TRMM Program. And I gave my talk, and I thought it went really well, and after the talk finished, a couple of really senior NASA program managers came up to me and told me I had just screwed up the whole thing! And I didn't even know why. They said they called Joanne, and this was a disgrace, and the whole program was going to be in trouble. And so, tail between my legs, I went back to Washington, DC where Joanne was and, right away, she met with me and said, "Chris you did nothing wrong." And then she asked me what I talked about, and I told her, and she said it was absolutely good. She picked up the phone and called the program managers and told them that I had done nothing wrong and they should take back what they said. Then after she did that—and this is the part I really won't forget—she sat me down and said, "Chris, this is what you really did wrong." And, she said, "You couldn't have possibly known this, but you now are talking about both science and politics, and the people in the politics world hear the same thing but they see something different, and you've just got to be careful that when you talk to them that you say what they want to hear." I thought great, OK, and I've never forgotten that. Every time my graduate students give a talk to an audience with management in attendance, I tell them this story, and they are always grateful to understand that it's not just science, but there's politics behind it when the money gets big enough. I'll never forget that about Joanne, to this day. It's too late now that she's passed away, but I would have *always* given my life for that woman because she backed me up when I really needed it. I'll never forget her."[22]

Chris delivered this final thought with tears in his eyes.

From her early collaborations with Riehl to her later collaborative team work, Joanne was a prolific mentor, both of individual junior scholars and as an initiator and leader of group projects. When asked about mentoring in her NASA exit interview, she replied: "Mentoring is one of the favorite roles that I have played for more than fifty years.

[21] Simpson Papers, 4.8. Simpson Symposium, February 9–13, 2003.
[22] Kummerow, "Chris Kummerow on Joanne Simpson and the politics of science," YouTube.

I will be pleased to continue on a more limited scale."[23] She accomplished her mentoring without formal relationships with graduate students. In a recent family tree of the academic community of tropical meteorology she is depicted without mentees. This is undoubtedly an artifact of the way the chart was constructed, which favored traditional professor–student genealogies.[24]

The Big Picture

Joanne's life course combined intellectual curiosity, professional upward mobility, resistance to societal constraints, and a personal quest that never ended. She was born shortly after Congress ratified the Nineteenth Amendment to the Constitution, granting women the right to vote and just after Margaret Sanger established the American Birth Control League, a precursor of today's Planned Parenthood. In 1923, the year of her birth, the National Women's Party introduced an Equal Rights Amendment to the US Senate, which aimed to provide for the legal equality of the sexes and prohibit discrimination on the basis of sex. Joanne's father introduced her to the aerial world. He had a front row seat for some famous aviation stories, including Charles Lindbergh's solo trans-Atlantic flight in 1927, Amelia Earhart's flight in 1932, and the tragedies of the Lindbergh kidnapping and Earhart's mysterious disappearance. Meanwhile, the stock market crash of 1929 and the Great Depression changed all of society. Joanne's political awareness and allegiances were formed at this time, especially during the New Deal presidency of Franklin Roosevelt, who appointed Frances Perkins as Secretary of Labor, the first woman to hold a cabinet-level position.

The main events shaping the larger society during Joanne's later high school and college years centered around the outbreak of World War II. During her first year in college, the US reinstated the military draft and instituted the massive Lend-Lease war production program. The declaration of war against Japan, Germany, and Italy in December of her sophomore year precipitated events that led to her visit to Rossby and her enrolment as a weather instructor in the war effort.

[23] Simpson Papers, 3.6. Interview transcript re: Simpson's exit interview from NASA, September 3, 2004.
[24] R. E. Hart and J. H. Cossuth, "A Family Tree of Tropical Meteorology's Academic Community and its Proposed Expansion," *Bulletin of the American Meteorological Society* 94 (2013): 1837–48.

She married one week after the D-Day invasion in 1944, became a mother two weeks after the Trinity A-bomb test, and received her master's degree in August 1945, just after the bombings of Hiroshima and Nagasaki.

The Cold War framed Joanne's postwar political opinions. In 1947, the year of her first visit to Woods Hole, President Harry Truman signed the Federal Employees Loyalty Program aimed at searching out "disloyal persons" in the US government. Heightened fears of communist infiltration and influence set the stage for prominent hearings by the House Un-American Activities Committee and Senator Joseph McCarthy. The Pacific Cloud Hunt of 1957 occurred several months before the Soviet launch of Sputnik 1 inaugurated both the space age and the space race. These events shaped Joanne's attitudes toward government secrecy and bureaucracy.

Her UCLA years were concurrent with the beginning of the Civil Rights and environmental movements, the Cuban missile crisis, and the arrival of US troops in Vietnam. During her years with the weather service and NOAA, the Supreme Court legalized contraceptive use by married couples, the National Organization for Women got its start, the Apollo 11 astronauts landed on the Moon, Congress passed the Clean Air Act, President Richard Nixon established the Environmental Protection Agency, and a Supreme Court ruling in Roe v. Wade legalized abortion. While Joanne was at NASA, in 1986, the Supreme Court ruled that sexual harassment is a form of illegal job discrimination. In 2010, the year of her death, the navy's ban on women in submarines ended. Joanne had broken the ban on women flying in military aircraft some six decades earlier.[25]

Joanne was clearly precocious. She was swimming by age 3, reading by 4, and sailing and flying by 6. She finished high school at 17, and was working in the war effort by 19. She married Victor Starr by age 21, but he was more of a mentor than a romantic partner; she became a mother and a postgraduate student at 22; and she earned her PhD degree at age 26. All three of her husbands were scientists, and she wished that her special friend, photographer Claude Ronne, could have been more of one. She was raised surrounded by music, ballet, books, and plays, and loved astrophysics and British history almost as much as clouds.

---

[25] Dreir, "Who and What Changed America?"

Cambridge was both cultured and turbulent; the cottage and marshes of Humarock peaceful and inviting. Chicago was gritty and lonely; Falmouth open (in more ways than one) and Bohemian, while Los Angeles and the cliffs above Malibu turned into the sites of her worst nightmares. Washington, DC reeked of too much bureaucracy, but Miami beckoned, for a time, due to its abundance of new challenges, cumulus clouds, and sailing adventures with Bob. Charlottesville was bucolic and lovely, but claustrophobic and horribly sexist; Greenbelt was not really so much a place to live as it was a place to work, hard, in her dream job at NASA. During her few remaining years of retirement living, her high-rise apartment in the Southwest Waterfront of Washington, DC at least looked out across the open water and skyscape that she loved so much.

Joanne had plenty of privileges and plenty of troubles as a child and throughout her career. She treasured books, clouds, the seashore, and nurturing her circle of young friends. She invested herself completely in her studies—at the Buckingham School, at the University of Chicago, and through the rest of her life. Her ambition knew no bounds. In response to the war emergency, she trained as a weather cadet at the tender age of 19 and was teaching recruits at age 20. When her unhappy first marriage disqualified her from teaching at Chicago, she supported herself and her son David as a jack-of-all-trades physics instructor at Illinois Institute of Technology. She turned her attention to clouds in Herb Riehl's seminar on tropical meteorology and earned a PhD degree under his supervision—the first granted to a woman in the United States. Woods Hole was a great place to fly into clouds and photograph them, accompanied by the first real love of her life, Claude Ronne. Joanne's reputation was bolstered by her computerized model of convection, the first of its kind, and her memorable hot towers hypothesis, jointly proposed with Riehl. She and her second spouse Willem broke the nepotism barrier as full professors at UCLA, but tenure, two incomes, and a fancy house in Malibu could not buy them happiness; neither could the arrival of baby Karen, at least not at that moment. Joanne found real happiness only after age 40, with her marriage to Bob Simpson, the second love of her life, and from the satisfaction that came through mentoring really great people. She loved experimenting on clouds and hurricanes, but chafed under the bureaucracy of

NOAA and its focus on operational weather modification. She hoped for a new start at the University of Virginia, but found the sexism there abhorrent. She was in her element at NASA, where she worked from 1979 to 2004. Recognition, honors, and awards, many of them firsts for women in meteorology, poured in. She broke down sexist barriers (or flew over them) and opened up new opportunities for women in meteorology. After a rough start in life, Joanne reached the top of her profession through a combination of hard work, talent, and determination. She reconciled with the past, with her insecurities, chronic migraines, and depression. Most significantly, she found true love—twice. All the sacrifice, pain, and exertion had indeed been worthwhile.

Bibliography

Archival Materials

Simpson, Joanne. "Papers, 1890–2010, MC 779." Schlesinger Library, Radcliffe Institute, Harvard University, Cambridge, MA.

Simpson, Joanne (Malkus). "Joanne Malkus Simpson Papers, 1951–1964." Woods Hole Oceanographic Institution, Data Library and Archives, Record Group Identifier: MC-37. Woods Hole, MA.

University of Chicago Library, Department of Special Collections. Department of Meteorology; Division of the Physical Sciences; Office of the President, Hutchins Administration; Records. Chicago, IL.

Original Videos Embedded in Digital Footnotes

Simpson, J. "Joanne Simpson Vignettes," nineteen videos. Recorded Sept. 2017–Jan. 2018. Playlist posted by Carol Lipshultz on YouTube, May–Aug. 2018. https://www.youtube.com/playlist?list=PLt6Ke-3HAyJRUbwtRrTsxgK73v4ESeUws

Anthes, R. "Rick Anthes on Joanne Simpson and Bob as Lovebirds." YouTube video. 2:16. Recorded on Jan. 8, 2018. Posted by Carol Lipshultz, May 25, 2018. https://www.youtube.com/watch?v=TBsNyOW6Gm4&index=4&list=PLt6Ke-3HAyJRUbwtRrTsxgK73v4ESeUws

Cotton, W. R. "Bill Cotton on Joanne Simpson, Why He Chose His Graduate School." YouTube video. 1:03. Recorded on Jan. 9, 2018. Posted by Carol Lipshultz, Aug. 6, 2018. https://www.youtube.com/watch?v=D9g4qWRWgHU&index=9&list=PLt6Ke-3HAyJRUbwtRrTsxgK73v4ESeUws

Halverson, J. "Jeff Halverson on Joanne Simpson and Working Together." YouTube video. 2:28. Recorded on Sept. 27, 2017. Posted by Carol Lipshultz, May 24, 2018. https://www.youtube.com/watch?v=V927U2r0G0g&index=8&list=PLt6Ke-3HAyJRUbwtRrTsxgK73v4ESeUws

Katsaros, K. "Kristina Katsaros on Joanne Simpson as Her Guardian Angel." YouTube video. 3:25. Recorded on Jan 10, 2018. Posted by Carol Lipshultz, July 30, 2018. https://www.youtube.com/watch?v=dx-PDQ8FkgM&index=12&list=PLt6Ke-3HAyJRUbwtRrTsxgK73v4ESeUws

Katsaros, K. "Kristina Katsaros on Joanne Simpson Cartoon." YouTube video. 1:52. Recorded on Jan 10, 2018. Posted by Carol Lipshultz, July 30, 2018. https://www.youtube.com/watch?v=Hzkqb8SxWr0&index=10&list=PLt6Ke-3HAyJRUbwtRrTsxgK73v4ESeUws

Kummerow, C. "Chris Kummerow on Joanne Simpson and the Politics of Science." YouTube video. 2:12. Recorded on Jan. 9, 2018. Posted by Carol Lipshultz, May 24, 2018. https://www.youtube.com/watch?v=8uhQE7sxlEo

LeMone, P. "Peggy LeMone on Joanne Simpson and CBS Conquest Program." YouTube video. 1:40. Recorded on Oct. 18, 2017. Posted by Carol Lipshultz, May 24, 2018. https://www.youtube.com/watch?v=_mEalM-bado&list=PLt6Ke-3HAyJRUbwtRrTsxgK73v4ESeUws&index=6

LeMone, P. "Peggy LeMone on Joanne Simpson and TOGA COARE." YouTube video. 1:11. Recorded on Oct. 18, 2017. Posted by Carol Lipshultz, July 25, 2018. https://www.youtube.com/watch?v=MknOhXzRUEI&list=PLt6Ke-3HAyJRUbwtRrTsxgK73v4ESeUws&index=3

LeMone, P. "Peggy LeMone on Joanne Simpson and Women in Meteorology." YouTube video. 2:18. Recorded on Oct. 18, 2017. Posted by Carol Lipshultz, Aug. 5, 2018. https://www.youtube.com/watch?v=9l2alEJ5Fi4&index=17&list=PLt6Ke-3HAyJRUbwtRrTsxgK73v4ESeUws

LeMone, P. "Peggy LeMone on Joanne Simpson NASA Retirement Song." YouTube video. 1:22. Recorded on Oct. 18, 2017. Posted by Carol Lipshultz, Aug. 5, 2018. https://www.youtube.com/watch?v=6H_kxJlBkWw&list=PLt6Ke-3HAyJRUbwtRrTsxgK73v4ESeUws&index=17&t=0s

Rutledge, S. "Steven Rutledge on Joanne Simpson." YouTube video. 2:49. Recorded on Jan. 9, 2018. Posted by Carol Lipshultz, Aug. 5, 2018. https://www.youtube.com/watch?v=qqSAia73_34&t=71s&list=PLt6Ke-3HAyJRUbwtRrTsxgK73v4ESeUws&index=6

Washington, W. "Warren Washington on Joanne Simpson and Climate Models." YouTube video. 1:13. Recorded on Oct. 18, 2017. Posted by Carol Lipshultz, Aug. 5, 2018. https://www.youtube.com/watch?v=RSis4NS-Pd4&list=PLt6Ke-3HAyJRUbwtRrTsxgK73v4ESeUws&index=14

References

"Agnew Reports to Nixon: Hurricane Warnings Must Be Improved," *Boston Globe* (Aug. 26, 1969), 1.

Anthes, R. "Remembering Joanne Simpson: The Life of a Legendary Meteorologist," *UCAR Magazine* (March 5, 2010). https://www2.ucar.edu/atmosnews/in-brief/1549/remembering-joanne-simpson

Atlas, D., G. Holland, and P. LeMone. "Obituaries: Joanne Simpson, 1923–2010," *Bulletin of the American Meteorological Society* 91 (2010): 938–9.

Atlas, D. and M. A. LeMone. "Joanne Simpson," *Memorial Tributes: National Academy of Engineering* 15 (2011): 368–75.

Atlas, D. *Reflections: A Memoir.* Boston: American Meteorological Society, 2001.

"Awards," American Meteorological Society. https://www.ametsoc.org/ams/index.cfm/about-ams/ams-awards-honors/awards

Bamzai, A.S. "The NSF's Role in Climate Science," *Oxford Research Encyclopedia of Climate Science*, 2020. DOI: 10.1093/acrefore/9780190228620.013.802

Barrett, A. *Women at Imperial College: Past, Present And Future.* World Scientific, 2017.

Beck, A. L. "An Application of the Principles of Bjerknes' Dynamic Meteorology in a Study of Synoptic Weather Maps for the United States." MA thesis, University of California at Berkeley, 1922.

Behl, M., L. Merner, and R. Pandya. "Diversity at AMS: Insights from the AMS Membership Survey," *Bulletin of the American Meteorological Society* 98 (2017): 1980–4.

Bengelsdorf, I. S. "Woman of the Year: Scientist With Her Feet on Cloud 9," *Los Angeles Times* (Dec 20, 1963), D1.

Börngen, M. "Lammert, Luise Charlotte." "Biografien historischer Frauen-persönlichkeiten: Wissenschaft Stadt." Leipzig, 2018. https://www.leipzig.de/jugend-familie-und-soziales/frauen/1000-jahre-leipzig-100-frauenportraets/detailseite-frauenportraets/projekt/lammert-luise-charlotte

Boyer, John W. *University of Chicago: A History*. Chicago: University of Chicago Press, 2015.

Braham, R. R., Jr. "Interview of Roscoe Braham, June 19–21, 2002," by Steven Cole. American Meteorological Society/University Corporation for Atmospheric Research, Tape-Recorded Interview Project.

Braham, R. R., Jr. and E. A. Neil. "Modification of Hurricanes Through Cloud Seeding," *National Hurricane Research Report* 16, Washington, DC: US Weather Bureau, 1958.

Byers, H. R. "The Founding of the Institute of Meteorology at the University of Chicago," *Bulletin of the American Meteorological Society* 57 (1976): 1343–5.

Cloud Systems, Hurricanes, and the Tropical Rainfall Measuring Mission (TRMM): A Tribute to Dr. Joanne Simpson, W.-K. Tao and R. F. Adler, eds. Boston: American Meteorological Society, *Meteorological Monographs* 29 (2003).

Conway, E. *Atmospheric Science at NASA: A History*. Baltimore: Johns Hopkins University Press, 2008.

Cotton, W. R. and R. A. Pielke, Sr. *Human Impacts on Weather and Climate*. Cambridge: Cambridge University Press, 1995.

DeMaria, M. "A History of Hurricane Forecasting for the Atlantic Basin, 1920–1995," in *Historical Essays on Meteorology, 1919–1995*, J. R. Fleming, ed. Boston: American Meteorological Society, 1996, pp. 263–305.

Des Jardins, J. *The Madame Curie Complex: The Hidden History of Women in Science*. New York: The Feminist Press at CUNY, 2010.

Dietsch, M. *Untersuchungen über die Änderung des Windes mit der Höhe in Zyklonen*. Leipzig: Metzger & Wittig, 1918.

Dorffel, K. "Die physiklische Arbeitsweise des Gallenkamp-Verdunstungsmessers und seine Anwendung auf mikroklimatische Fragen." PhD dissertation, Leipzig, 1935.

Downie, L., Jr. "New Agency a Step Toward Weather Control," *Washington Post* (Aug 21, 1965), B1.

Dreir, P. "Who and What Changed America? A 20th Century Timeline." https://www.huffingtonpost.com/peter-dreier/150-moments-that-changed-_b_7513366.html

Dry, S. *Waters of the World*. Chicago: University of Chicago Press, 2019.

Dunn, G. E. "Cyclogenesis in the Tropical Atlantic." *Bulletin of the American Meteorological Society* 21 (1940): 215–29.

Dunn, G. E. "Tropical Cyclones," in *Compendium of Meteorology*, T. Malone, ed. Boston: American Meteorological Society, 1951, pp. 887–901.

"Engineer and the Weatherman," *Illinois Tech Engineer* (Oct. 1948).

"First Lady of Tropical Meteorology," *Discovery Earth* (c.2010). https://www.youtube.com/watch?v=IGzcubQ8OpE.

Fleagle, R. G. Review of *The Tropical Atmosphere. Cloud Structure and Distributions over the Tropical Pacific Ocean*, J. S. Malkus and H. Riehl. University of California Press, Berkeley, 1964, *Science* n.s. 148 (1965): 809–10.

Fleming, J. R. "Carl-Gustaf Rossby: Theorist, Institution Builder, Bon Vivant," *Physics Today* 70 (2017): 50–6.

Fleming, J. R. *Fixing the Sky: The Checkered History of Weather and Climate Control.* New York: Columbia University Press, 2010.

Fleming, J. R. "Global Environmental Change and the History of Science," in *Cambridge History of Science*, vol. 5, *The Modern Physical and Mathematical Sciences*, M. J. Nye, ed. Cambridge: Cambridge University Press, 2003, pp. 634–50.

Fleming, J. R. *Inventing Atmospheric Science: Bjerknes, Rossby, Wexler and the Foundations of Modern Meteorology.* Cambridge, MA: MIT Press, 2016.

Friehe, C. M. and H. M. Stommel. "Andrew F. Bunker: Pioneering in Air–Sea Interaction Research 1946–79," *Bulletin of the American Meteorological Society* 72 (1991): 44–9.

"GATE, History of the GARP Atlantic Tropical Experiment." https://www.ametsoc.org/sloan/gate.

Gentry, R. C. "Hurricane Modification," in *Weather and Climate Modification*, W. N. Hess, ed. New York: Wiley, 1974, pp. 497–521.

Gerould, J. [Profile]. *The Packet of the Buckingham School.* Cambridge, MA, June, 1940.

Grimes, A. "Equatorial Meteorology," in *Compendium of Meteorology*, T. Malone, ed. Boston: American Meteorological Society, 1951, pp. 881–6.

Hardaker, Paul. "Dr. Joanne Simpson," *Weather* (Sept. 24, 2010). https://doi.org/10.1002/wea.667.

Harper, K. C. *Make it Rain: State Control of the Atmosphere in Twentieth-Century America.* Chicago: University of Chicago Press, 2017.

Hart, R. E. and J. H. Cossuth, "A Family Tree of Tropical Meteorology's Academic Community and its Proposed Expansion," *Bulletin of the American Meteorological Society* 94 (2013): 1837–48.

Hartten, L. M. and M. LeMone. "The Evolution and Current State of the Atmospheric Sciences 'Pipeline'." *Bulletin of the American Meteorological Society* 91 (2010): 942–56.

Haurwitz, B. "The Height of Tropical Cyclones and of the 'Eye' of the Storm," *Monthly Weather Review* 63 (1935): 45–9.

Hewitt, N. A. *No Permanent Waves: Recasting Histories of U.S. Feminism.* New Brunswick, NJ: Rutgers University Press, 2010.

Hinkel, L. "Celebrating Pauline (Polly) Morrow Austin: A Founder of Radar Meteorology" (Jan. 26, 2017). https://eapsweb.mit.edu/news/2017/celebrating-pauline-morrow-austin-founder-radar-meteorology

Hirschberg, C. "My Mother, the Scientist," *Popular Science*, April 18, 2002.

Hitchfeld, W. F. "The National Hail Research Experiment," Review of *Hailstorms of the Central High Plains* by C. A. Knight and P. Squires. *Science* 219 (1983) 379–80.

Holland, G. "Remembering Joanne Simpson: The Life of a Legendary Meteorologist," *UCAR Magazine* (March 5, 2010). https://www2.ucar.edu/atmosnews/in-brief/1549/remembering-joanne-simpson

Houze, R. A., Jr. "Joanne Simpson (1923–2010)," *Nature* 464 (2010): 696.

Hurricane Camille: A Report to the Administrator. Washington, DC: US Department of Commerce, ESSA, Sept. 1969.

Hurricane! Coping with Disaster: Progress and Challenges Since Galveston, 1900, R. H. Simpson, R. A. Anthes, and M. Garstang, eds. Washington, DC: American Geophysical Union, 2003.

"Into the Eye of a Cyclone," Townsville, Australia, *Advertiser* (Feb. 18, 1993), 6.

Jamison, K. R. "Manic-Depressive Illness and Creativity," *Scientific American* 272 (Feb. 1995); 62–7.

Krishnamurti, T. N., L. Stefanova, and V. Misra. *Tropical Meteorology: An Introduction.* New York: Springer, 2013.

Kuettner, J. P. "General Description and Central Program of GATE," *Bulletin of the American Meteorological Society* 55 (1974): 712–19.

Kummerow, C., W. Barnes, T. Kozu, J. Shiue, and J. Simpson. "The Tropical Rainfall Measuring Mission (TRMM) Sensor Package," *Journal of Atmospheric and Oceanic Technology* 15 (1998): 809–17.

Kundsin, R. B. *Women and Success: The Anatomy of Achievement.* New York: Morrow, 1973.

Lammert, L. C. *Der mittlere Zustand der Atmosphäre bei Südföhn.* Leipzig: Metzger & Wittig, 1920.

Langmuir, I. "The Production of Rain by a Chain Reaction in Cumulus Clouds at Temperatures Above Freezing," *Journal of Meteorology* 5 (1948): 175–92.

Langwell, P. A. "Inhomogeneities of Turbulence, Temperature, and Moisture in the West Indies Trade-Wind Region," *Journal of Meteorology* 5 (1948): 243–6.

LeMone, M. A. "What We Have Learned about Field Programs," in *Cloud Systems, Hurricanes, and the Tropical Rainfall Measuring Mission (TRMM): A Tribute to Dr. Joanne Simpson,* W.-K. Tao and R. F. Adler, eds. Boston: American Meteorological Society, *Meteorological Monographs* 29 (2003), pp. 25–35.

LeMone, P. "Remembering Joanne Simpson: The Life of a Legendary Meteorologist," *UCAR Magazine* (March 5, 2010). https://www2.ucar.edu/atmosnews/in-brief/1549/remembering-joanne-simpson

LeMone, P. "What Really Happened in Honiara…" Photocopied booklet, n.p., 1993.

Lettau, H. [and Käte Dorffel Lettau]. "Interview with Heinz Lettau, March 10, 2002," by Sharon Nicholson. American Meteorological Society/University Corporation for Atmospheric Research, Tape-Recorded Interview Project.

Lewis, J. M. "WAVES" Forecasters in World War II (with a Brief Survey of Other Women Meteorologists in World War II)," *Bulletin of the American Meteorological Society* 76 (1995): 2187–202.

Lewis, J. M., M. G. Fearon, and H. E. Klieforth. "Herbert Riehl, Intrepid and Enigmatic Scholar," *Bulletin of the American Meteorological Society* 93 (2012): 963–85.

Malkus-Benjamin, Karen. "Video Interview of Karen Malkus-Benjamin, Jan. 30, 2018," by Carol Lipshultz. Author's personal files.

Malkus, J.S. "Aeroplane Studies of Trade-Wind Meteorology," *Weather* 8 (1953): 291–9.

Malkus, J.S. "Certain Features of Undisturbed and Disturbed Weather in the Trade-Wind Region," PhD dissertation, Dept. of Meteorology, University of Chicago (1949), 54 pp.

Malkus, J.S. "Cumulus, Thermals, and Wind," *Boating* 13 (1949): 6–9, 12.

Malkus, J.S. "The Effects of a Large Island upon the Trade-Wind Air Stream," *Quarterly Journal of the Royal Meteorological Society* 81 (1955): 538–50.

Malkus, J.S. "Effects of Wind Shear on Some Aspects of Convection," *Transactions of the American Geophysical Union* 30 (1949): 19–25;

Malkus, J.S. "Large Scale Interactions," *The Sea: Ideas and Observations on Progress in the Study of the Seas*, vol. 1. New York and London: Interscience Publishers, 1962, pp. 88–294

Malkus, J.S. "On the Maintenance of the Trade Winds," *Tellus* 8 (1956): 335–50.

Malkus, J.S. "On the Structure and Maintenance of the Mature Hurricane Eye," *Journal of Meteorology* 15 (1958): 337–49.

Malkus, J.S. "The Origin of Hurricanes," *Scientific American* 197 (Aug. 1957): 33–9.

[Malkus, J.S.]. "Origin of Weather," CBS Television Network, 16-mm film, sound, black and white, 26 min., 1960.

Malkus, J.S. "Pacific Cloud Hunt: An Episode Behind the Scenes in Weather Research." *University of Chicago Magazine* 50 (May 1958), 4–10, 24.

Malkus, J.S. "Recent Advances in the Study of Convective Clouds and Their Interaction with the Environment," *Tellus* 4 (1952): 71–87.

Malkus, J.S. "The Slopes of Cumulus Clouds in Relation to External Wind Shear," *Quarterly Journal of the Royal Meteorological Society* 78 (1952): 530–42.

Malkus, J.S. "Some Results of a Trade-cumulus Cloud Investigation," *Journal of Meteorology* 11 (1954): 220–37.

Malkus, J.S. "Trade-Wind Clouds," *Scientific American* 189 (Nov. 1953): 31–5.

Malkus, J.S. "Tropical Convection: Progress and Outlook," in *Proceedings of the Symposium on Tropical Meteorology, Rotorua, New Zealand*. Wellington: New Zealand Meteorological Service, 1964, pp. 247–77.

Malkus, J.S. "Tropical Weather Disturbances—Why Do So Few Become Hurricanes?" *Weather* 13 (1958): 75–89.

Malkus, J.S. and A. Bunker. "Observational Studies of the Air Flow over Nantucket Island During the Summer of 1950," *Papers in Physical Oceanography and Meteorology* 12 (1952).

Malkus, J.S., A.F. Bunker, and K. McCasland. "Observational Studies of Convection," *Soaring* 14 (1950): 13–15.

Malkus, J.S., A.P. Bunker, B. Haurwitz, and H. Stommel. "Vertical Distribution of Temperature and Humidity over the Caribbean Sea," *Papers in Physical Oceanography and Meteorology* 11 (1949).

Malkus, J.S. and H.R Riehl, *Cloud Structure and Distributions over the Tropical Pacific Ocean.* Berkeley: University of California Press, 1964.

Malkus, J.S. and H.R. Riehl. "Cloud Structure and Distributions Over the Tropical Pacific Ocean," *Tellus* 16 (1964): 275–87.

Malkus, J.S. and H.R. Riehl. "On the Dynamics and Energy Transformations in Steady State Hurricanes," *Tellus* 12 (1960): 1–20.

Malkus, J.S. and H.R. Riehl. "Some Aspects of Hurricane Daisy, 1958," *Tellus* 13 (1961): 181–213.

Malkus, J.S., C. Ronne, and M. Chaffe. "Cloud Patterns in Hurricane Daisy, 1958," *Tellus,* 13 (1961): 8–30.

Malkus, J.S. and R.H. Simpson. "Modification Experiments on Tropical Cumulus Clouds," *Science* 145 (1964): 541–8.

Malkus, J.S. and R.H. Simpson. "Note on the Potentialities of Cumulonimbus and Hurricane Seeding Experiments," *Journal of Applied Meteorology* 3 (1964): 470–5.

Malkus, J.S. and R.H. Simpson. "The Ocean as a Laboratory for Weather Modification Experiments," *Oceanus* 11 (1964): 16–24.

Malkus, J.S. and M. Stern. "The Flow of a Stable Atmosphere Over a Heated Island," Parts I and II. *Journal of Meteorology* 10 (1953): 30–41, 105–120.

Malkus, J.S. and R.T. Williams. "On the Interaction Between Severe Storms and Large Cumulus Clouds," *Meteorological Monographs* 5 (1963): 59–63.

Malkus, J.S. and G. Witt. "The Evolution of a Convective Element: A Numerical Calculation," in *The Atmosphere and the Sea in Motion,* B. Bolin, ed. Rockefeller Institute, Oxford University Press, 1959, pp. 425–39.

Murakami, T. "Scientific objectives of the Monsoon Experiment (MONEX)," *GeoJournal* 3 (1979): 117–36.

National Hurricane Center. "Hurricanes in History," http://www.nhc.noaa.gov/outreach/history

Noble, I. "Joanne Simpson, Meteorologist," in *Contemporary Women Scientists of America.* New York: Julian Messner, 1979, pp. 31–48.

North, G. R. "GATE and TRMM," in *Cloud Systems, Hurricanes, and the Tropical Rainfall Measuring Mission (TRMM): A Tribute to Dr. Joanne Simpson,* W.-K. Tao and R. F. Adler, eds. Boston: American Meteorological Society, *Meteorological Monographs* 29 (2003), pp. 201–6.

Oreskes, N. "Stommel, Henry Melson," in *Complete Dictionary of Scientific Biography* 24. Detroit: Charles Scribner's Sons, 2008, pp. 527–32.

Palmer, C. E. "Reviews of Modern Meteorology, 5. Tropical Meteorology," *Quarterly Journal of the Royal Meteorological Society* 78 (1952): 126–64.

Palmer, C. E. "Tropical Meteorology," in *Compendium of Meteorology*, T. Malone, ed. Boston: American Meteorological Society, 1951, pp. 859–80.

Pielke, R. A., Sr. "Joanne Simpson—An Ideal Model of Mentorship." *AMS Meteorological Monographs* 29 (2003): 17–24.

Ponte, L. "Alien Ice," in *Iceberg Utilization*, A. A. Husseiny, ed. New York: Pergamon Press, 1978, pp. 11–19.

Popkin, R. *The Environmental Science Services Administration*. New York: Praeger, 1967.

Priestly, C. H. B. "William Christopher Swinbank 1913–1973," *Records of the Australian Academy of Science* 3 (1974).

Quinn, S. *Marie Curie: A Life*. Boston: Da Capo Press, 1996.

"RAMS Description," http://rams.atmos.colostate.edu/rams-description.html.

Raymond, C. A. and B. Carlson. "My Daughter the Scientist," *University of Chicago Magazine* 78 (March 1986), 15–17, 24, 29.

Reiter, E. "Herb: Personal Recollections by Elmar R. Reiter, with the Help of Others," *Meteorology and Atmospheric Physics* 67 (1998): 5–14.

"Remembering Joanne Simpson: The Life of a Legendary Meteorologist," *UCAR Magazine* (March 5, 2010). https://www2.ucar.edu/atmosnews/in-brief/1549/remembering-joanne-simpson

Riehl, H. R. "Aerology of Tropical Storms," in *Compendium of Meteorology*, T. Malone, ed. Boston: American Meteorological Society, 1951, pp. 902–13.

Riehl, H. R. "Interview with Herbert Riehl, Sept. 9, 1989," by Joanne Simpson. American Meteorological Society/University Corporation for Atmospheric Research, Tape-Recorded Interview Project.

Riehl, H. R. *Tropical Meteorology*. New York; McGraw-Hill, 1954.

Riehl, H. R. and J. S. Malkus. "On the Heat Balance and Maintenance of Circulation in the Trades," *Quarterly Journal of the Royal Meteorological Society* 83 (1957): 21–8.

Riehl, H. R. and J. S. Malkus. "On the Heat Balance in the Equatorial Trough Zone," *Geophysica* 6 (1958): 503–35.

Riehl, H. R., J. S. Malkus, T. C. Yeh, and N. E. LaSeur. "The North-East Trade of the Pacific Ocean," *Quarterly Journal of the Royal Meteorological Society* 77 (1951): 598–626.

Riehl, H. R. and J. Simpson. "The Heat Balance of the Equatorial Trough Zone, Revisited," *Contributions to Atmospheric Physics* 52 (1979): 287–304.

Rossiter, M. W. *Women Scientists in America: Before Affirmative Action, 1940–1972*. Baltimore: Johns Hopkins University Press, 1995.

Rossiter, M. W. *Women Scientists in America: Forging a New World Since 1972*. Baltimore: Johns Hopkins University Press, 2012.

Rossiter, M. W. *Women Scientists in America: Struggles and Strategies to 1940*. Baltimore: Johns Hopkins University Press, 1992.

"Science of Superstorms: Playing God with the Weather." London: British Broadcasting Corporation/Discovery Channel, 2007.

"Seeding of Storm to be Tried Today: Weathermen hope to break up hurricane in Atlantic," *New York Times* (Sept. 1, 1965), 14.

Sherer, A. "Who's Who in GATE," (1999). https://www.ametsoc.org/sloan/gate/index.html

Simpson, J. "AMS Responds to Challenges: Outgoing President's Report by Joanne Simpson, at the Annual Business meeting of the American Meteorological Society, Anaheim, California, 4 February 1990," *Bulletin of the American Meteorological Society* 71 (1990): 942–4.

Simpson, J. "The GATE Aircraft Program: A Personal View," *Bulletin of the American Meteorological Society* 57 (1976): 27–30.

Simpson, J. "Iceberg Utilization: Comparison with Cloud Seeding and Potential Weather Impacts," in *Iceberg Utilization*, A. A. Husseiny, ed. New York: Pergamon Press, 1978, pp. 624–39.

Simpson, J. "Interview of Joanne Simpson, Sept. 6, 1989," by Margaret LeMone. American Meteorological Society/University Corporation for Atmospheric Research, Tape-Recorded Interview Project.

Simpson, J. "Meteorologist," in *Successful Women in the Sciences: An Analysis of Determinants,* R. B. Kundsin, ed. *Annals of the New York Academy of Sciences* 208 (1973): 41–6.

Simpson, J. "On Some Aspects of Sea–Air Interaction in Middle Latitudes," *Deep Sea Research* 16 (1969): 233–61

Simpson, J. "Oral History Interview with Joanne Simpson, August 2, 2000," by Kristine Harper. Interview conducted at NASA Goddard, Greenbelt, MD, and available from the American Institute of Physics.

Simpson, J. "Papers of Joanne Simpson, 1890–2010 (inclusive), 1950–1995 (bulk): A Finding Aid." Arthur and Elizabeth Schlesinger Library on the History of Women in America. Radcliffe Institute for Advanced Study, Harvard University, Cambridge, MA., January 2014.

Simpson, J. "TRMM (Tropical Rainfall Measuring Mission) Data Set Potential in Climate Controversy," https://pielkeclimatesci.wordpress.com/2008/02/27/trmm-tropical-rainfall-measuring-mission-data-set-potential-in-climate-controversy-by-joanne-simpson-private-citizen/

Simpson, J. "A Satellite Mission to Measure Tropical Rainfall." NASA, 1988.

Simpson, J. "Women in the Atmospheric Sciences. Astounding Progress since World War II. Personal Viewpoint of Joanne Simpson in 2002." *WeatherZine* 34 (June 2002). http://sciencepolicy.colorado.edu/zine/archives/34/editorial.html

Simpson, J., R. Adler, and G. North. "A Proposed Tropical Rainfall Measuring Mission (TRMM) Satellite," *Bulletin of the American Meteorological Society* 69 (1988): 278–95.

Simpson, J. and B. Brown. "Potential of Summer Rain Augmentation by Cloud Seeding in the Mid-Atlantic states" *Virginia Journal of Science* 29 (1978): 146–56.

Simpson, J. and C. Kummerow. "The Tropical Rainfall Measuring Mission and Vern Suomi's Vital Role," *10th Conference on Atmospheric Radiation: A Symposium* with *Tributes to the Work of Verner E. Suomi, 28 June–2 July 1999*. Boston: American Meteorological Society, 1999, pp. 44–51.

Simpson, J., C. Kummerow, W.-K. Tao, and R. F. Adler. "On the Tropical Rainfall Measuring Mission (TRMM)." *Meteorology and Atmospheric Physics* 60 (1996): 19–36.

Simpson, J. and M. LeMone. "Women in Meteorology," *Bulletin of the American Meteorological Society* 55 (1974): 122–131.

Simpson, J., R. H. Simpson, D. A. Andrews, and M. A. Eaton. "Experimental Cumulus Dynamics," *Reviews of Geophysics* 3 (1965): 387–431.

Simpson, J., R.H. Simpson, J.R. Stinson, and J.W. Kidd. "Stormfury Cumulus Experiments: Preliminary results 1965," *Journal of Applied Meteorology* 5 (1966): 521–5.

Simpson, J. and V. Wiggert. "Models of Precipitating Cumulus Towers," *Monthly Weather Review* 97 (1969): 471–89.

Simpson, R. H. "Exploring the Eye of Typhoon Marge 1951," *Bulletin of the American Meteorological Society* 33 (1952): 286–98.

Simpson, R. H. *Hurricane Pioneer: Memoirs of Bob Simpson*. Boston: American Meteorological Society, 2014.

Simpson, R. H. "Interview of Robert H. Simpson, September 6, 1989." by Edward Zipser. American Meteorological Society/University Corporation for Atmospheric Research, Tape-Recorded Interview Project.

Simpson, R. H., N. E. LaSeur, R. C. Gentry, L. F. Hubert, and C. L. Jordan, "Objectives and Basic Design of the National Hurricane Research Project," Report no. 1. Washington, DC: US Department of Commerce, Weather Bureau, 1956.

Simpson, R. H. and J. S. Malkus. "An Experiment in Hurricane Modification: Preliminary Results," *Science* 142 (1963): 489.

Simpson, R. H. and J. S. Malkus. "Experiments in Hurricane Modification," *Scientific American* 211 (June 1964): 27–37.

Starr, J. G. "Note on Internal Oscillations," *Journal of Meteorology* 2 (1945): 120–2.

Starr, J. G. "Women in Meteorology," in *Weather Horizons,* F. D. Ashley, ed. Boston: American Meteorological Society, 1947, pp. 23–9.

Stellmacher, C. "Über den Einfluss von Luftdruck und Wind auf Hoch- und Niedrigwasser an der deutschen Ostseeküste," PhD dissertation, Münster, 1920.

Stommel, H. "Entrainment of Air into a Cumulus Cloud," *Journal of Meteorology* 4 (1947): 91–4.

Sullivan, P. "Joanne Malkus Simpson, World-Renowned Atmospheric Scientist." Obituary. *Washington Post* (March 8, 2010), B4.

Tao, W.-K., R. Adler, S. Braun, F. Einaudi, B. Ferrier, J. Halverson, G. Heymsfield, C. Kummerow, A. Negri, and R. Kakar. "Summary of a Symposium on Cloud Systems, Hurricanes and TRMM: Celebration of Dr. Joanne Simpson's

Career—The First 50 years," *Bulletin of the American Meteorological Society* 81 (2000): 2463–74.

Tao, W-K., J. Chern, R. Atlas, D. Randall, M. Khairoutdinov, J. Li, D. E. Waliser, A. Hou, X. Lin, C. Peters-Lidard, W. Lau, J. Jiang, and J. Simpson. "A Multiscale Modeling System: Developments, Applications, and Critical Issues," *Bulletin of the American Meteorological Society* 90 (2009): 515–34.

Tao, W.-K., J. Halverson, M. LeMone, R. Adler, M. Garstang, R. Houze, Jr., R. Pielke, Sr., and W. Woodley. "The Research of Dr. Joanne Simpson: Fifty Years Investigating Hurricanes, Tropical Clouds, and Cloud Systems," in *Cloud Systems, Hurricanes, and the Tropical Rainfall Measuring Mission (TRMM): A Tribute to Dr. Joanne Simpson,* W.-K. Tao and R. F. Adler, eds. Boston: American Meteorological Society, *Meteorological Monographs* 29 (2003), pp. 1–16.

Theon, J. S. "My View of the Early History of TRMM and Dr. Joanne Simpson's Key Role in Winning Mission Approval," in *Cloud Systems, Hurricanes, and the Tropical Rainfall Measuring Mission (TRMM): A Tribute to Dr. Joanne Simpson,* W.-K. Tao and R. F. Adler, eds. Boston: American Meteorological Society, *Meteorological Monographs* 29 (2003), pp. 175–80.

"Tranquilizer Prepared for Stalled Betsy," *Washington Post* (Sept. 1, 1965), A3.

Turner, R. D. "Teaching the Weather Cadet Generation: Aviation, Pedagogy, and Aspirations to a Universal Meteorology in America, 1920–1950," in *Intimate Universality: Local and Global Themes in the History of Weather and Climate,* J. R. Fleming, V. Jankovic, and D. R. Coen, eds. Sagamore Beach, MA: Science History Publications, 2006, pp. 141–73.

Uccellini, L. "Interview of Louis Uccellini, Oct. 19, 2016," by James Fleming and Maya Meltsner, NOAA Headquarters, Silver Spring, MD. Author's personal files.

University of Chicago, *Daily Maroon,* various, 1940–1942. http://campub.lib. uchicago.edu/search/?keyword=Joanne+Gerould

"University of Chicago, Department of Meteorology," in *Weather Horizons,* F. D. Ashley, ed. Boston: American Meteorological Society, 1947, pp. 44–9.

US Department of Commerce, National Oceanic and Atmospheric Administration, *Project Stormfury.* Washington, DC: GPO, 1977.

US Department of Labor. "The Outlook for Women in Geology, Geography, and Meteorology," *Bulletin of the Women's Bureau* No. 223–7. Washington, DC: USGPO, 1948.

US National Oceanic and Atmospheric Administration. "Hurricane Research Division History." http://www.aoml.noaa.gov/hrd/hrd_sub/beginning.html

US National Research Council. *Earth Observations from Space: The First Fifty Years of Scientific Achievements.* Washington, DC: National Academies Press, 2008.

US Navy. Bureau of Naval Weapons, Research, Development, Test and Evaluation Group, Meteorological Management Group. "Technical Area Plan for Weather Modification and Control." TAP No. FA-4. January 1, 1965.

Weier, J. "Joanne Simpson (1923–2010)," https://earthobservatory.nasa.gov/
   Features/Simpson/

Wexler, H. "Structure of Hurricanes as Determined by Radar," *Annals of the
   New York Academy of Sciences* 48 (1946–1947): 821–44.

White, R. M. "The Making of NOAA, 1963–2005," *History of Meteorology* 3 (2006):
   55–63.

"Willem Malkus, professor emeritus of mathematics, dies at 92," http://news.
   mit.edu/2016/willem-malkus-professor-emeritus-mathematics-dies-0608

Willoughby, H. E., D. P. Jorgensen, R. A. Black, S. L. Rosenthal. "Project
   STORMFURY: A Scientific Chronicle 1962–1983," *Bulletin of the American
   Meteorological Society* 66 (1985): 505–14.

Woodcock, A. H. and J. Wyman. "Convective Motion in Air Over the Sea,"
   Annals of the New York Academy of Sciences 48 (1947): 749–76.

Woodley, W. L. "The Effect of Airborne Silver Iodide Pyrotechnic Seeding on
   the Dynamics and Precipitation of Supercooled Tropical Cumulus Clouds."
   PhD dissertation, Florida State University, 1969.

Woodley, W. L. "Precipitation Results from a Pyrotechnic Cumulus Seeding
   Experiment, *Journal of Applied Meteorology* 9 (1970): 242–57.

Index